ELECTRICAL AND TELECOMMUNICATIONS
TECHNICIANS SERIES
EDITORS: A. TRANTER, S. F. SMITH

Electronics for technicians

Electronics
for technicians

Roger Hamilton
Wolverhampton Polytechnic

Oxford   New York   Toronto   Melbourne
OXFORD UNIVERSITY PRESS
1979

Oxford University Press, Walton Street, Oxford OX2 6DP

OXFORD LONDON GLASGOW
NEW YORK TORONTO MELBOURNE WELLINGTON
KUALA LUMPUR SINGAPORE JAKARTA HONG KONG TOKYO
DELHI BOMBAY CALCUTTA MADRAS KARACHI
NAIROBI DAR ES SALAAM CAPE TOWN

Published in the United States
by Oxford University Press, New York

British Library Cataloguing in Publication Data

Hamilton, Roger
    Electronics for technicians. – (Electrical and
    telecommunications technicians series).
    1. Electronic apparatus and appliances
    I. Title   II. Series
    621.381   TK7870   79–40151

ISBN 0-19-859353-8
ISBN 0-19-859354-6 Pbk

Printed in Great Britain by
Fletcher & Son Ltd, Norwich

Contents

Preface

As will be explained in the Introduction, this book is intended for students on certain T.E.C. certificate and diploma courses. For the first time in a technician course electronics is treated as a subject in itself, rather than as an offshoot of electrical engineering.

I should like to thank numerous colleagues and friends for their help and advice. In particular my family, for their tolerance, David Weighton for his comments on the layout, and my daughter Ruth for preparing some of the drawings.

Wolverhampton, 1977   R.H.

1 Introduction

1.1 Electronics

This book is intended for students taking courses leading to a T.E.C. Certificate requiring a knowledge of electronics at levels II and III. Such a student will have studied, or be exempt from, a basic vocational unit such as 'Telecommunications systems' or 'Electronic systems and practice'. These units deal with the applications of electronics in various areas.

Both the above-mentioned units deal with the communication of information and the handling of signals which carry information. Information needs to be transmitted efficiently from one place to another for various reasons. There are commercial reasons, e.g. to transmit urgent business information to any part of the world instantaneously; there is entertainment—radio, television, etc.; and there is telemetry—the transmission of measured values from a place normally inaccessible such as a satellite or space station. Closely allied to these applications is the subject of 'Radar and navigational aids'—a topic on its own.

Industrial electronics is a subject that some students will study in addition to 'pure' electronics. It is concerned with the use of electronics in industry, and, as well as the transmission of information and measurements referred to above, it also covers such topics as machine-tool control, and the control of heavy-current devices such as motors and generators. Under this heading we may also include rectification (see Chapter 3).

Yet another application of electronics is in the computer. Whilst a detailed study of computers is well beyond the scope of this book, a start is made in Chapter 8 on logic.

Electronics was in its infancy before the Second World War. The explosion in the subject since then has been quite extraordinary. Much of this book might be considered out of date already. Nevertheless, equipment designed using the techniques to be described will be in use for many years to come.

1.2 Layout of this book

This volume is primarily designed to cover T.E.C. Standard Units 76/010 (Electronics level II) and 76/009 (Electronics level III). These are part of the Certificate in Telecommunications and will also form part of a number of certificates in Electronics. However, some colleges might use other syllabuses and therefore some topics in this book go slightly beyond the level of the two units mentioned above.

In some areas it is clear which topics are level II and which are level III. In certain places the two levels merge, and differentiation is difficult. Briefly the chapters and levels are as follows.

Chapter 2	§§2.1–2.6 Level II
	§2.7 Level III
	§2.8 Levels II and III
Chapter 3	Level II; §3.6 is not in 76/010 or 76/009
Chapter 4	This is the most difficult chapter to divide into levels. §§4.1 and 4.2 form an introduction, although the topic 'noise' is in level III.
	§§4.3 and 4.4 are level II topics, but include some fundamental work not in 76/010 or 76/009.
	§4.5 Level II
	§4.6 Level III
	§4.7 Not really in 76/010 or 76/009 but work which is very important. You may omit it (but consult your lecturer).
Chapters 5 and 6	Level III
Chapter 7	The basic concept of waveforms and the simple LC oscillator are level II. The rest is level III.
Chapter 8	§§8.1–8.6 Level II
	§§8.7 and 8.8 are not in 76/010 or 76/009 but serve as an introduction to further logic (such as is contained in 76/011—Digital techniques).
Chapter 9	The CRT is in level II.

It is essential that you read the above breakdown in consultation with your lecturer. Try to obtain a copy of the syllabus being used at your College.

While studying at level II you should read those areas indicated above, attempting the relevant test questions. When studying at level III you can, of course, omit chapters such as 3 and 8 but try to reread those where the levels are mixed, such as Chapter 4.

Lastly, do attempt all the test questions. They occasionally introduce new ideas, and may thus be regarded as part of the text.

A note on valves (chapter 9) Apart from certain high-power applications and some modern hi-fi circuits, virtually no designs are being undertaken nowadays using valves (apart, of course, from the cathode-ray tube (CRT)). Valves were included in T.E.C. syllabuses at the request of industrial representatives. How important they are to you, or your employer, will determine your attitude to them. Again, consult your lecturer. It is essential to cover such topics as thermionic emission to deal with the CRT.

1.3 Relationship to other subjects

It is expected that you have studied, or are exempt from, the following units: (a) Physical Science I 75/004, particularly §D Electricity and §I Further Electricity; (b) Mathematics I 75/005; and (c) Electronic Systems and Practice or Telecommunications Systems 76/007.

During level II you should be studying: (a) Mathematics II 75/033 (or equivalent) and (b) Electrical Principles II 75/019 (or equivalent). Whilst the whole of the second unit is important, the following sections are particularly so for your study of electronics: A1, Units; B2, Circuits; C3, Capacitors and capacitance, particularly §3.12; E, Electromagnetic induction, particularly §§5.8–5.11; F, Alternating voltages and currents; G, Single-phase a.c. circuits, particularly reactance and impedance calculations; and H, Measuring instruments and measurements.

During level III you will be studying Electrical Principles III 75/010 (or equivalent). Of particular importance to electronics are the following sections: A, A.C. circuits; C, D.C. transients; D, Single-phase transformers; and F, Measuring instruments and measurements.

This is not meant to imply that other sections are not important, but only that they are not needed in studying 76/010 or 76/009. Clearly, as an example, 75/010 section E (D.C. machines) is essential if you work in the field of industrial electronics (now or in the future).

2 Semiconductor devices

2.1 Conductors, insulators, and semiconductors

To understand the way in which semiconductor diodes and transistors operate it is necessary to know something of the structure of matter. If the following very elementary treatment of atomic structure gives some idea of the working of semiconductor devices it will have achieved its object.

The atom may be thought of as a positively charged nucleus or core surrounded by a number of electrons moving round it in orbits, rather like the planets circling round the sun. The nucleus is composed of two types of particle, protons and neutrons, both having the same mass ($1 \cdot 67 \times 10^{-27}$ kg), the proton having a charge of $+1 \cdot 60 \times 10^{-19}$ C and the neutron being uncharged. The electrons circling around the nucleus each have a charge of $-1 \cdot 60 \times 10^{-19}$ C and, in a neutral atom, the number of electrons equals the number of protons. As the electrons are very much lighter than protons or neutrons ($9 \cdot 11 \times 10^{-31}$ kg) the mass of the atom is almost entirely situated in the nucleus. We shall, however, only be concerned here with the orbiting electrons, for they determine the electrical properties.

The electrons arrange themselves in certain definite orbits, usually referred to as shells, and are labelled by the physicist as the K, L, M, etc. shells, the K shell being the innermost. The *rules* of shell filling are too complicated to be considered here in detail, and all we need consider is the fact that any particular atom consists of a number of full shells and an outer shell which may or may not be full.

The simplest atom is hydrogen with just one electron in the K shell. Helium has two K electrons and this is, in fact, the maximum number allowed in the K shell. Lithium, the next element, has two K electrons and one in the L shell, which can accommodate eight. The number of electrons in the outermost shell of an atom is the number of *valence* electrons. The valence electrons have a considerable effect on both the chemical and physical properties of a substance.

Some elements, such as copper, have just one electron in their outermost shells, and this electron requires very little energy to remove it from its orbit. In fact, at room temperature, all these outer electrons have been given enough thermal energy to detach them from the atoms, and in a piece of such material they will be moving about at random. If a battery is connected across a rod of solid copper, for example, electrons will drift towards the positive terminal, and this drift of charge is what we call an electric current. It is surprising to note that a typical value for the velocity of this drift might be in the order of 10^{-6} m/s (the actual value depends on a number of factors) and this means that an electron might take nearly 12 days to travel along a one-metre bar! (Of course the movement, and hence the current, starts as soon as the battery is connected.) It must be remembered that by convention we call it a flow of current from positive to negative (electric currents were found to exist before the electron was discovered!). Substances like copper are known as *conductors* and have an important property in that their resistivity increases with an increase in temperature, because the moving electrons are impeded by the vibrating atoms.

Other substances have valence electrons which are very tightly bound to the nucleus, particularly if the outermost shell is full. There are then hardly any free conduction electrons and the substance is called an *insulator*.

Substances range between the extremes of excellent conductors and excellent insulators. We are particularly concerned with a group of elements with four valence electrons (tetravalent materials) which are known as *semiconductors*. The two of main interest to us are silicon with 14 electrons arranged 2-8-4 and germanium with 32 electrons arranged 2-8-18-4. To put matters into perspective, the resistivity of copper at 300 K is $10^{-8}\,\Omega\,$m and that of mica, a good insulator, $10^{10}\,\Omega\,$m. Silicon and germanium have resistivities of 2300 $\Omega\,$m and 0·45 $\Omega\,$m respectively at this temperature. However, apart from this somewhat higher resistivity, conduction in a semiconductor is different from that in a conductor in two very important ways, as will be seen by considering just how a semiconductor passes current.

A crystal of germanium or silicon is formed by the joining together of atoms, each atom attaching itself by its four valence electrons to four neighbours, building up a three-dimensional structure as shown in Fig. 2.1. Figure 2.2 shows a two-dimensional representation of it, the two lines between atoms representing the two valence electrons, one from each atom, holding the structure together.

At room temperature some of the electrons will have sufficient energy to break away from this structure, becoming free to move about. They will leave behind a region in the structure which is positively charged. These positive regions are known as holes. The material as a whole, however, has no charge as there must always be an equal number of holes and electrons. Not only do the free electrons move about at random, as in a conductor,

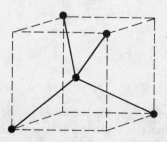

Fig. 2.1. Arrangement of atoms in a semiconductor.

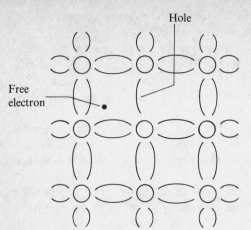

Fig. 2.2. *Two-dimensional representation of a semiconductor.*

but the holes also move about by a process of electrons moving into the hole, leaving behind a hole elsewhere. It is rather like a row of cinema seats, all occupied except for the one at the end. If the person in seat two moves to the empty seat and the person in seat three moves to seat two, and so on, we could describe the action in simpler terms by saying that the empty seat is moving along the row. It is convenient to think of the hole moving as though it were an actual positive particle. Of course, some electrons and holes will annihilate themselves but, in equilibrium, there will be as many new hole–electron pairs being formed, so that the total number will remain constant at a particular temperature.

If a battery is connected across a rod of semiconductor material not only do electrons drift towards the positive terminal but holes drift towards the negative one. If the battery is connected to the rod by copper wires, only electrons can flow in the copper; holes reaching the negative end of the rod will be filled by electrons entering the rod from the copper.

Thus conduction in pure, or *intrinsic*, semiconductor is very different from conduction in a conductor such as copper. This is the first important difference referred to above. The second difference is the temperature effect: if the temperature of a pure semiconductor rises, more hole-electron pairs are formed and the resistivity decreases. Hence a semiconductor has a negative temperature coefficient of resistance, unlike metals, which have a positive one. In fact at extremely low temperatures a pure semiconductor would be a good insulator. This negative temperature coefficient is very important, and will have quite an effect on the way transistor circuits are designed.

2.2 n- and p-type semiconductors

Conduction in an intrinsic semiconductor has been seen to consist of the movement of *equal* numbers of holes and electrons. One or other type of *carrier* may be made to predominate by the addition of impurities in very small controlled amounts.

n-type   n-type material is made by the addition of minute quantities of substances such as arsenic, antimony, or phosphorus which have five valence electrons, that is which are *pentavalent*. The material is still substantially tetravalent and the impurity atoms are forced into the tetravalent structure, as shown in Fig. 2.3. The numbers in the circles are the number of valence electrons of the atoms. The extra electron from the pentavelent atom behaves rather like the single electron in a copper atom, and at room temperature will move about the material at random. The atom left behind will carry a positive charge, although the material as a whole is uncharged. The positively charged atom, or *ion*, is, of course, fixed in the structure and unable to move about. The substance formed is a much better conductor than a pure semiconductor; the addition of impurity atoms in the ratio of 1 in 10^8 reduces the resistivity nearly 100 times. Conduction is now mainly due to electron movement and electrons are therefore called the *majority* carriers. The impurity atoms, because they give free electrons to the material, are known as *donors*. There will still be some holes due to the thermal breakdown of bonds and these are called *minority* carriers. Because current is mainly carried by negative electrons the material is known as n-type semiconductor.

p-type   Trivalent impurities such as indium, boron, and aluminium are added to pure semiconductor to form p-type materials. Having only three valence electrons the impurity atoms can only attach themselves to three of their four neighbours, as shown in Fig. 2.4. The hole so formed in the structure is in need of an electron, and it will attract one from a nearby semiconductor atom, thus producing a positive hole there, which will then move further about the crystal at random as explained earlier. The impurity atom will have gained an electron and carry a negative charge. The material

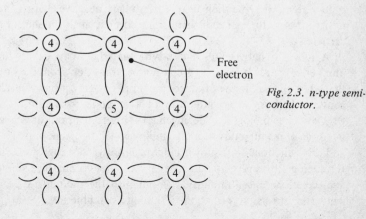

Free electron

Fig. 2.3. n-type semiconductor.

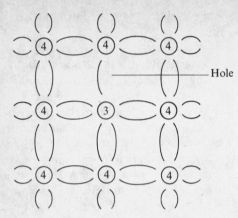

Fig. 2.4. *p-type semiconductor*

now conducts mainly due to positive carriers and is called a p-type semi-conductor. Holes are the *majority* carriers and electrons the minority ones. The impurity is called an *acceptor* impurity because it accepts electrons. As with an n-type semiconductor, the resistivity is much lower than that of intrinsic material.

2.3 The p–n junction

Like intrinsic semiconductors p- and n-type are resistive materials, that is, if the applied voltage is reversed in polarity the same current will flow in the opposite direction. As we shall see the p–n junction has rectifying properties: it can pass current more easily in one direction than the other. As the junction is fundamental to all semiconductor devices it is important to try to follow the explanation of its properties.

It is convenient to consider n-type material as consisting of stationary positively charged atoms (the impurity atoms) and a number of mobile electrons. The stationary semiconductor atoms, much more numerous than the impurity ones, do not enter into this explanation. Similarly p-type material may be thought of as stationary negative charges and mobile positive holes. In each case there will also be a small number of minority carriers.

A junction between p- and n-type materials may be produced in a number of ways. It must be a continuous transition from the one type to the other and cannot be produced by simply placing pieces of each type in contact. It is not the intention of this book to consider diode and transistor technology in detail, but a brief mention of some of the methods of manufacture that are employed will be given.

Early junctions were grown junctions. A crystal was grown in liquid of, for example, n-type material. After it was about a quarter of an inch long a piece of acceptor impurity was put into the molten n-type material, converting it to a p-type and thus forming a p–n junction.

The three most common methods in use today are alloying, diffusion, and epitaxy.

An *alloy* junction is made by taking a piece of n-type germanium (melting point 950°C) and placing on it a small piece of the trivalent substance indium (melting point 156°C). The temperature is then raised well above 156°C an the molten indium dissolves some germanium, forming p-type material on cooling. The p- and n-type materials then have leads attached and the whole is encapsulated, usually in the atmosphere of an inert gas. A similar method is used with silicon, employing aluminium as the impurity.

Diffused junctions are made by heating one type of semiconductor to near its melting point in a gaseous atmosphere consisting of the other type of impurity. The impurity diffuses a small distance into the material, changing its type and forming a p-n junction. For silicon the impurities used are often either phosphorous (as donor) or boron (as acceptor).

An *epitaxial* junction consists of a silicon substrate onto which is grown a crystal of p- or n-type material by heating it (to over 1000°C) in a gaseous atmosphere of silicon trichloride and hydrogen together with the relevant impurity. The hydrogen combines with the chlorine, and silicon (p- or n-type as the case might be) is deposited onto the substrate. The impurity can then be changed, to produce a p-n junction.

Whichever process is used to produce the junction it is convenient to imagine materials of the two types being brought together. The carriers in each piece are moving about at random, and some of the majority carriers situated near the junction will diffuse into the other material, where they will meet a high concentration of the opposite carriers, many of them recombining and producing a region on either side of the junction with very few carriers, known as the depletion layer (Fig. 2.5). This depletion layer, which will be of the order of 0·5 μm thick, behaves rather like a piece of insulating material situated between the p- and n-types of semiconductor. As will be seen in Fig. 2.5, a potential exists across the junction due to the stationary impurity atoms, the n-side being positive with respect to the p-side. This potential is known as a *barrier* potential and

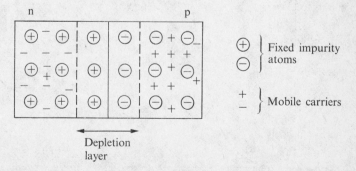

Fig. 2.5. The p–n junction (no bias).

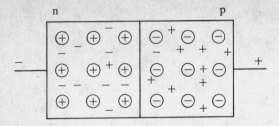

Fig. 2.6. The p–n junction (forward bias).

opposes any further movement of majority carriers across the junction. A contact potential of this kind exists whenever two dissimilar metals are in intimate contact. However we may complete the external circuit between the ends of the block shown, the sum of all the contact potentials round the circuit would just equal zero, some being positive and some negative. This explains why the junction potential cannot be used to drive current round an external circuit.

Forward voltage

Figure 2.6 shows the junction with an applied voltage across it such that the p-type material is positive with respect to the n-type. The applied voltage opposes the contact potential, lowering the potential barrier. If it is large enough, holes and electrons will flow easily, from the p- and n-regions respectively, across the junction. The current so produced rises very rapidly as the applied voltage is increased. Figure 2.7 shows the *forward* characteristics for both germanium and silicon junctions. As can be seen a forward-biased silicon junction carrying a few milliamperes has a voltage drop of 0·6 to 0·7 V across it, a germanium one somewhat less.

Reverse voltage

If the polarity of the applied voltage is reversed (Fig. 2.8), a very different

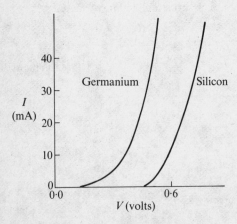

Fig. 2.7. Forward characteristics of silicon and germanium junctions.

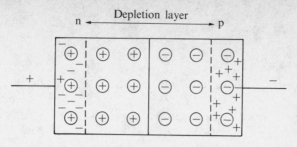

Fig. 2.8. The p–n junction (reverse bias).

situation exists. The applied voltage is now in the same sense as the contact potential, leading to an increase in the potential barrier, opposing the flow of majority carriers. There will, however, be a small flow of current due to the minority carriers, holes from the n-region to the right and electrons from the p-region to the left. At room temperature this *reverse* current is very small, being typically a few microamperes for germanium and less than a microampere for silicon, and it is almost independent of the value of the applied reverse voltage. However, owing to the increase in the thermal breakdown of the semiconductor bonds with temperature, an increase in temperature produces a large increase in reverse current. In modern devices a 10 °C rise roughly doubles the current.

Reverse characteristics of silicon and germanium junctions are shown in Fig. 2.9. If the reverse voltage increases too far the current rapidly increases, probably destroying the device. This increase is due to the *avalanche* effect produced by the minority carriers moving so fast, because of the large voltage, that in colliding with atoms they break down bonds, producing more hole–electron pairs and hence more current. In normal operation, of course, this *reverse breakdown voltage* must be avoided, but the avalanche effect is used in some practical devices, such as some zener diodes.

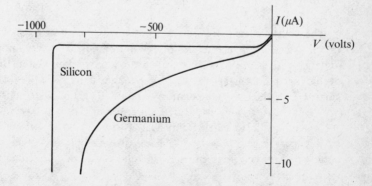

Fig. 2.9. Reverse characteristics of semiconductor junctions.

Anode ———————▷|— Cathode   *Fig. 2.10. Symbol for a semiconductor junctions.*

2.4. The junction diode

The junction diode, as already indicated, consists of a junction, with suitable leads attached, sealed in a container in an inert atmosphere. In devices intended to carry large currents the junction will be bonded to a heat sink, probably many times larger than the device itself, the purpose of which is to remove heat from the junction.

The junction diode has many uses in electrical engineering. Small *signal diodes* are used in communication engineering for such purposes as the demodulation of radio signals and large diodes are used for rectification, or the conversion of a.c. to d.c., which will be the subject of Chapter 3.

Signal diodes may be in the form of a cylinder a few millimetres long and perhaps 0·1 mm in diameter. A diode to carry 10 A may be 20–30 mm long and 10–20 mm in diameter and be intended to be mounted on a metal heat sink. Such a diode carrying 1 A and dropping, say, 1 V will dissipate 1 W.

The circuit symbol used for a diode is given in fig. 2.10, the p-material being the *anode*, that is the terminal which must be the more positive to forward bias the device and cause a large current to flow. In many diode applications the current which flows when the anode is less positive than the *cathode* (the reverse current) may be neglected.

Diode resistance

The resistance of a device is the voltage across it divided by the current flowing, and in this sense a forward-biased diode has a resistance. However, the resistance is not constant, but alters with the applied voltage. Figure 2.11 shows the forward characteristic of a silicon diode. The resistance at point A is given by OC divided by OD, that is the reciprocal of the slope of the line OA. Similarly the resistance at B is the reciprocal of the slope of OB and, as OB is steeper than OA, the resistance at B is less than that at A.

The resistance just referred to is called the static or d.c. resistance of the diode, to distinguish it from a second kind of resistance known as the *dynamic resistance*, sometimes called the a.c. resistance, slope resistance, or incremental resistance, r. Let us assume that the diode is passing a current OD, that is it is operating at point A on the characteristic. If we now try to change the current by a small amount, that is move up or down the characteristic to a new point near A, the device will offer a resistance to this *change* given by the reciprocal of the slope of the tangent to the curve at A. As will be seen, the dynamic resistance is less than the static resistance.

It can be shown that the dynamic resistance in ohms of a junction

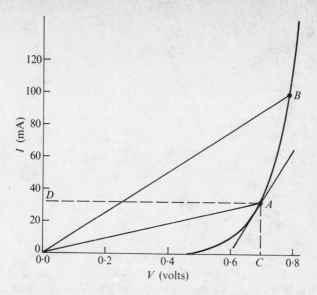

Fig. 2.11. *Static and dynamic resistance of a diode.*

diode at room temperature is given, very approximately, by 26 mV divided by the current flowing in milliamperes. Thus at 1 mA the resistance is 26 Ω, at 2 mA 13 Ω, etc. Hence

(2.1)
$$r(\Omega) = \frac{26}{I}$$

where I is expressed in mA.

Load line

A common problem is to determine the current which will flow if a diode and resistor, or *load*, are connected in series to a given d.c. voltage (Fig. 2.12).

The voltage across the resistor R_L is IR_L; hence, by Kirchhoff's Laws, the voltage across the diode, V_D, is given by

(2.2)
$$V_D = V - IR_L.$$

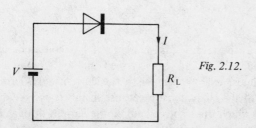

Fig. 2.12.

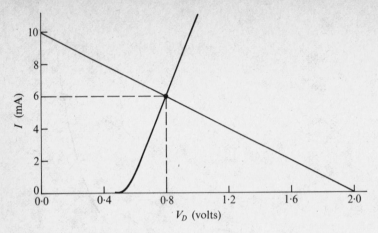

Fig. 2.13.

This equation can be plotted on a graph whose axes are I and V_D. These are the same axes as those used for plotting the static characteristic of the diode and the intersection of the two graphs is the operating point of the diode.

As it is a straight line, known as the load line, it can be plotted from two points, the easiest choice being the two places where it crosses the axes. It will cross the V_D-axis when I is zero, that is when

(2.3) $V_D = V.$

It will cross the I-axis when V_D is zero, or when

(2.4) $I = \dfrac{V}{R_L}.$

As an example, Fig. 2.13 shows a load line on a diode characteristic for $V = 2\,\text{V}$ and $R_L = 200\,\Omega$. The load line crosses the voltage axis at 2 V (see eqn. (2.3)), and the current axis at $2\,\text{V}/200\,\Omega = 10\,\text{mA}$ (see eqn (2.4)). Hence at the operating point I can be seen to be about 6 mA and the diode voltage about 0·8 V leaving 1·2 V across R_L. (Note that if a current of 6 mA flows in a resistance of 200 Ω the voltage across it is 1·2 V.) This constructional technique is very important in electronics and will be used again when dealing with transistors and valves.

2.5 The zener diode

The breakdown in the reverse characteristic of a silicon diode occurs at a very well-defined voltage and, as was seen in Fig. 2.9, the curve after breakdown is almost exactly parallel to the current axis. This is a very interesting part of the characteristic because in this region the diode has a voltage across it which does not depend on the current through it, at least over a large range of currents. It can thus be used in this region of its characteristic as a voltage reference, rather like, although not quite as good as, a standard cell.

Special diodes, known as *zener diodes*, are manufactured with breakdown voltages much lower than those of normal diodes. The name comes from the fact that at voltages below about 6 V the breakdown is not due to the avalanche effect but to the so-called zener effect, the name zener diode being used, however, even for diodes with breakdowns above 6 V.

Zener diodes are made with breakdowns, or zener voltages, using the same numerals as preferred resistance values, e.g. 5·6 V, 6·8 V. They are available ranging from a volt or so up to 20 V or more. One important application of them is in voltage stabilizer circuits, as will be seen in Chapter 3.

The circuit symbol for a zener diode is given in Fig. 2.14. It must be realized that a zener diode is usually operated with current flowing in a reverse direction, that is the cathode will be positive with respect to the anode. It will need to be operated in series with a resistor to limit the diode current to a safe value.

Anode   Cathode   *Fig. 2.14. Symbol for a zener diode.*

2.A Test questions

1. The negatively charged particles in the atom are
   - (a) protons
   - (b) neutrons
   - (c) electrons
   - (d) positrons

2. Silicon and germanium are
   - (a) tetravalent
   - (b) trivalent
   - (c) pentavalent

3. The conductivity of silicon and germanium
   - (a) decreases with an increase in temperature
   - (b) increases with an increase in temperature
   - (c) is constant with temperature

4. A rod of pure silicon connected to a battery will pass a current which is the same, but in the opposite direction, if the polarity of the battery is reversed. This is
   - (a) true
   - (b) false
   - (c) not sufficient information to answer

5. In intrinsic germanium
   - (a) holes are the majority carriers
   - (b) electrons are the majority carriers
   - (c) holes and electrons exist in equal numbers

6. A semiconductor has a temperature coefficient of resistance that is
   (a) negative
   (b) positive

7. Arsenic, antimony, and phosphorous are all
   (a) pentavalent
   (b) tetravalent
   (c) trivalent

8. Pentavalent impurities are known as
   (a) acceptor impurities
   (b) donor impurities

9. A p-type semiconductor has
   (a) pentavalent impurities
   (b) tetravalent impurities
   (c) trivalent impurities

10. An impurity ratio of 1 in 10^8
    (a) increases the resistivity about 100 times
    (b) decreases the resistivity about 100 times
    (c) decreases the resistivity about 10^8 times
    (d) has no significant effect on the resistivity
    (e) decreases the conductivity about 100 times

11. Free electrons in p-type material
    (a) are the majority carriers
    (b) take no part in conduction
    (c) are the minority carriers
    (d) exist in the same numbers as holes

12. A p–n junction is formed by placing very flat surfaces of p- and n-type material together. This is
    (a) true
    (b) false
    (c) only one of the methods which may be used

13. An unbiased p–n junction has a depletion layer whose thickness is of the order of
    (a) $0.5\,\mu m$
    (b) $0.005\,\mu m$
    (c) $50\,\mu m$
    (d) $10^{-12}\,m$

14. A potential exists across an unbiased junction such that
    (a) the p-side is positive with respect to the n-side
    (b) the p-side is negative with respect to the n-side

15. A forward-biased junction has an applied potential with
    (a) the positive to the p-side
    (b) the negative to the p-side

16. In a circuit the voltage across a silicon diode is found to be $0.7\,V$ with the anode positive with respect to the cathode. The diode is
    (a) reverse biased
    (b) forward biased
    (c) faulty

17. Repeat problem 16 if the reading is 10 V.

18. The reverse current in a diode
    (a) is constant with temperature
    (b) doubles with a 10 °C rise in temperature
    (c) doubles with a 10 °C fall in temperature

19. The dynamic resistance of a diode is less than its static resistance.
    (a) true
    (b) false
    (c) it depends on the temperature

20. The dynamic resistance of a diode, at room temperature, passing a forward current of 5 mA will be approximately
    (a) 26 Ω (c) 130 Ω
    (b) 5 Ω (d) 26 mΩ

21. A resistor of value R_L is in series with a diode and a battery of voltage V. To find the diode current, I, and voltage, V_D, a load line is drawn on the diode characteristic. The intercept on the current axis will be
    (a) $V - IR_L$ (c) $V \cdot R_L$
    (b) V/R_L (d) $26\,\text{mV}/R_L$

22. A zener diode is usually operated in the breakdown region.
    (a) true
    (b) false

23. The breakdown in a zener diode is due to the zener effect and not to the avalanche effect.
    (a) true
    (b) false
    (c) it depends on the breakdown voltage

24. An important application of the zener diode is as a rectifier.
    (a) true
    (b) false

2.6 The junction transistor

The junction transistor is basically a *sandwich* of either p-type semiconductor between two pieces of n-type (an n–p–n transistor) or n-type between two pieces of p-type (a p–n–p transistor). In either case it consists of two p–n junctions and may be manufactured by any of the methods used for diodes. Thus a p–n–p alloy junction transistor can be made by repeating the process referred to above on the other side of the n-type slice and letting the indium dissolve just enough germanium to leave a thin layer of n-type between the two p-regions (Fig. 2.15).

The emitter, base, and collector are also shown in Fig. 2.15. It may be seen that the collector is of greater area than the emitter and, in fact, in large transistors it is physically connected to a metal heat sink. This is because, as will be seen, the carriers (holes or electrons) travel to the collector, and on reaching it give up energy which appears in the form of

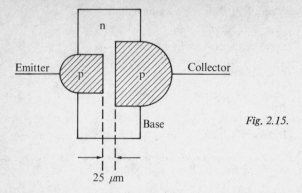

Fig. 2.15.

heat. The base region is extremely thin, being typically 25 μm in an alloy transistor, and considerably less in planar types.

The circuit symbols for both p–n–p and n–p–n transistors are given in Fig. 2.16. In both types the emitter carries an arrow-head and by convention the arrow of any p–n junction points from the p-type material to the n-type. (This is the direction of conventicial current flow when forward biased.)

In *normal* operation, as in an amplifier, the emitter-base junction is kept forward biased, the collector–base junction being reverse biased.

Common base configuration

Consider an n–p–n transistor biased as shown in Fig. 2.17. Firstly note that if V_{EB} is zero, V_{CB} will cause only a small reverse, or leakage, current to flow round the right-hand loop of the circuit. It will be remembered from the study of the diode that this very small current increases rapidly as the temperature is raised (it roughly doubles for a 10 °C rise). In a transistor it is given the symbol I_{CB0}, (pronounced I, C, B, nought) which means the collector current (I_C) for a common-base circuit (which will be explained below) with zero emitter current.

Now if V_{EB} is raised above zero a current will flow across the emitter-base junction. This current will consist of both electrons from emitter to base and holes from base to emitter, although the concentrations of impurities in the emitter and base regions are adjusted, during manufacture, such that the large majority of this current (over 99 per cent) will consist of electrons from emitter to base. Because of the thinness of the base region, most of these electrons will enter the depletion layer of the reverse-

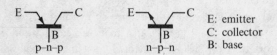

Fig. 2.16. Symbols for transistors.

E: emitter
C: collector
B: base

biased base–collector junction. Here they will be accelerated owing to the density gradient in the depleted region and enter the collector region where, as majority carriers, they will flow to the collector terminal. In a normal transistor well over 99 per cent of the electrons from the emitter will flow out of the collector terminal, the remainder flowing out of the base terminal.

The fraction of the emitter current which leaves at the collector is a very important parameter of a transistor, known as the short-circuit common-base forward current gain, α (alpha). It is the current gain when the collector current is flowing in a short circuit. The gain will vary with the load connected in the collector circuit, but a short circuit is a convenient 'load' to use to define α. α is thus given by:

(2.5)
$$\alpha = \frac{\text{Change in } I_C}{\text{Change in } I_E} \quad \text{with } V_{CB} \text{ constant.}$$

It should be noted that the change in I_C is slightly less than that in I_E and α will therefore be less than one. As the input is between emitter and base and the output between collector and base, the base is common to input and output: hence the name common-base configuration.

α is normally above $0 \cdot 99$ and can be as high as $0 \cdot 998$. It is not constant but varies with emitter current, collector voltage, and temperature. If α is, say, $0 \cdot 99$ then an emitter current of $1 \, \text{mA}$ would lead to a collector current of $0 \cdot 99 \, \text{mA}$ and hence a base current of $1 - 0 \cdot 99 = 0 \cdot 01 \, \text{mA}$. However when we use the transistor these three currents will remain in these ratios only as long as conditions are such that α remains $0 \cdot 99$.

If the voltage V_{CB}, Fig. 2.17, is now varied, it is found that the collector current, I_C, for a given emitter current, I_E, remains almost constant, increasing only very slightly as V_{CB} is increased. In fact, even with V_{CB} zero most of the emitter current still flows to the collector and V_{CB} has to be reversed in polarity to reduce I_C to zero.

Output characteristics

The circuit of Fig. 2.18 can be used to obtain values of I_C as V_{CB} is varied for different fixed values of I_E. I_E is set to a suitable value by adjusting R_1, and readings of I_C for different values of V_{CB} (set by R_2) are obtained. It will be necessary to reverse the polarity of V_2 to reduce I_C to zero, and great care must be taken, otherwise I_C will rapidly increase in the *reverse direction* and the transistor could burn out. Sets of readings are taken for a range of values of I_E (including zero) and the results, when plotted, are known as the *output characteristics*, because we are plotting the *output voltage* (V_{CB}) against the *output current* (I_C). Figure 2.19 shows a typical set of output characteristics for a small n–p–n transistor. It should be noted, during the experiment, that

(2.6)
$$I_E = I_C + I_B.$$

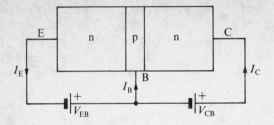

Fig. 2.17.

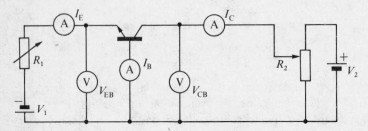

Fig. 2.18.

This is Kirchhoff's first law applied to the transistor.

The fact that the output characteristics are almost horizontal means that the resistance of the transistor output to *changes* in current (the dynamic output resistance) is very high. It should be quite clear from

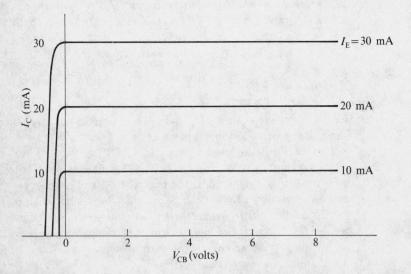

Fig. 2.19. Common-base output characteristics (n–p–n transistor).

Fig. 2.19 that a large change in voltage produces hardly any change in current. The dynamic output resistance is the reciprocal of the slope of the output characteristics and will be of the order of many megohms.

Input characteristics

The output characteristics are not, of course, the only graphs that may be plotted using the circuit of Fig. 2.18. We are also concerned about what is happening at the input of the transistor, and this is explained by drawing the *input characteristic*. This is a graph of I_E against V_{EB} for a given value of V_{CB} (Fig. 2.20), and is very similar to the forward-biased diode characteristic, the voltage across the junction being of the order of 0·6 V. (The input characteristic will vary very little if V_{CB} is changed, so that it is usual to only show one such curve.)

Clearly, for given values of V_1 and R_1 (Fig. 2.18) the exact values of V_{EB} and I_E could be found by a load-line technique as for the diode, but such accuracy is rarely required in practice.

The dynamic input resistance may be found as the reciprocal of the slope of the input characteristic at the operating point, as in Fig. 2.11 for the diode. In practice the input resistance is given by $r = 26/I_E$, where I_E is measured in mA.

Transfer characteristic

As the input to the common-base transistor is the emitter current and the output is the collector current, a plot of these two currents is known as the *transfer characteristic*. Such a characteristic is shown in Fig. 2.21.

It will be seen that it is a straight line, not quite through the origin,

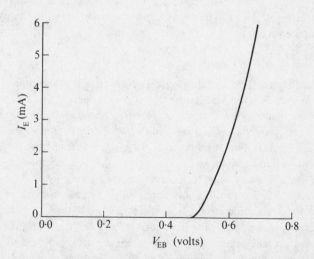

Fig. 2.20. Common-base input characteristic (n–p–n transistor).

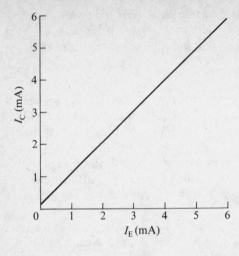

Fig. 2.21. The common-base transfer characteristic.

the intercept on the I_C-axis being I_{CBO}. The slope of the line is the value of α for the transistor. Whilst this characteristic is very important, there is not much need to plot it because, being straight, it can be specified by the two parameters I_{CBO} and α.

p–n–p transistors

Although we have considered only the n–p–n transistor in the preceding paragraphs, similar results apply to a p–n–p device. In this case, however, most of the current between emitter and base will consist of holes flowing from the emitter and most of them will enter the collector region to become majority carriers, as do the electrons in the n–p–n transistor.

When taking the characteristics of the p–n–p transistor the circuit shown in Fig. 2.18 can still be used, but the polarities of the two batteries will require reversing.

Common-base amplifier

Despite the fact that the short-circuit current gain is less than unity, the common-base circuit will act as an amplifier because, as will be shown below, a change of voltage on the emitter (input) will produce a much larger change of voltage across a load resistor connected in the collector (output) circuit. Since there is a very high output resistance (many megohms) a load resistor of a few kilohms will have no appreciable effect on the current in the collector circuit.

Figure 2.22 shows the connections of a common-base amplifier. It is not a very practicable circuit because it uses two batteries, which is inconvenient. V_1 and V_2 provide the correct bias for the two junctions, R_1 limits the current in the forward-biased emitter–base junction to a suitable value. If the junction is silicon it drops about 0·6 V and the emitter current is given by

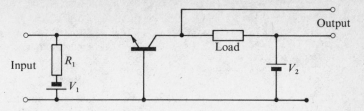

Fig. 2.22. The common-base amplifier.

$$(2,7) \qquad I_E = \frac{V_1 - 0 \cdot 6}{R_1}.$$

As has been explained above, the dynamic resistance of a forward-biased junction is very low. Let us assume that the emitter–base junction is operated in this example such that its dynamic resistance is $50\,\Omega$. If the input consists of a *change* in voltage at the emitter of, say, $1\,\text{mV}$, the *change* in emitter current produced will be $1\,\text{mV}/50\,\Omega = 0 \cdot 02\,\text{mA}$ and this will produce a *change* in collector current of about $0 \cdot 02 \times \alpha\,\text{mA}$. If α is $0 \cdot 99$ this change in collector current will be $0 \cdot 0198\,\text{mA}$. Now, because of the very high dynamic output resistance of the device, the load resistance could be as high as, say, $10\,\text{k}\Omega$ with little effect on the collector current. In this case the output voltage change, which is produced by the change in collector current flowing through the load resistor, is $0 \cdot 0198\,\text{mA} \times 10\,\text{k}\Omega = 0 \cdot 198\,\text{V}$ or $198\,\text{mV}$. Thus an input change of $1\,\text{mV}$ has produced an output change of $198\,\text{mV}$, a voltage gain of 198.

It is worth giving thought to the fact that the changing input voltage is feeding into a low resistance (in this example $50\,\Omega$) but that the current so produced (or 99 per cent of it) is flowing in a high resistance $(10\,\text{k}\Omega)$. It is this transfer of resistance that gives the transistor its name (being short for transfer resistor).

Common-emitter amplifier

It was emphasized above that however we use the transistor as long as the junctions are biased in the ways described, the three currents will be in the ratios of $1 : \alpha : (1 - \alpha)$ for $I_E : I_C : I_B$. It is perhaps easier to appreciate the significance of the following remarks if we use numbers rather than letters, and let us, as illustration, take α as $0 \cdot 99$ (although higher values are common with modern transistors). In this case the ratios are $1 : 0 \cdot 99 : 0 \cdot 01$. If we regard the base current as the input instead of the emitter current and still consider the collector current as the output we shall get a considerable current gain, in this case $0 \cdot 99/0 \cdot 01 = 99$. This circuit configuration is known as a *common-emitter circuit*, as the emitter is now common to both input and output, Fig. 2.23, and is today the most widely used configuration.

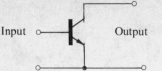

Input   Output

Fig. 2.23. The common-emitter configuration.

If we apply an input voltage between base and emitter, the base-current changes will obviously be much smaller than the corresponding emitter-current changes; in other words, because the base current is now the input current, the input resistance is much higher than it was when we were considering the emitter current as input, being typically a few kilohms. For reasons which will not be clear at present, it is usually given the symbol h_{ie}. A common-emitter input characteristic is shown in Fig. 2.24, and this, together with the output and transfer characteristics, can be plotted from data obtained using the circuit of Fig. 2.25. R_2 and R_3 can be used to alter the base and collector currents respectively. R_1 limits the base current to a safe value if R_2 is adjusted to zero.

A typical set of output characteristics is shown in Fig. 2.26. Two important differences between these curves and those for the common-base circuit will be seen. First the collector current reduces to zero without reversing the collector voltage (which is now V_{CE} and not V_{CB}, however) and secondly the characteristics are sloping upwards, meaning that the output resistance is less (of the order of tens of kilohms rather than megohms). Again for reasons not yet obvious, a symbol h_{oe} is used. It is in fact the output *conductance* and hence an output resistance of, say, 50 kΩ, which is typical, would be quoted as being an h_{oe} of 1/50 mS or 20 μS.

As well as the high current gain of this circuit there is a second advantage when compared with the common-base configuration, shown by a comparison of the input and output resistances. In the early days of transistors the common-base circuit was the only circuit in wide use. This was

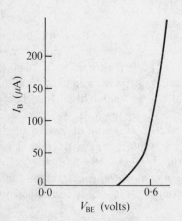

Fig. 2.24. A common-emitter input characteristic.

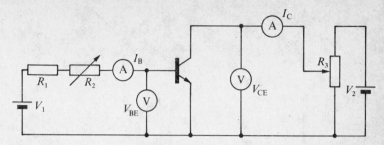

Fig. 2.25.

because early transistors had a very poor frequency response, and any transistor will always have a better frequency response when operating in the common-base configuration. Thus for an amplifier to operate even at relatively low frequencies it was necessary to use a common-base circuit. Unfortunately the very low input resistance of this circuit tended to *short out* the very high output resistance of the previous stage and it was necessary to use transformer coupling, with all the associated disadvantages, not least the cost. The input and output resistances of common-

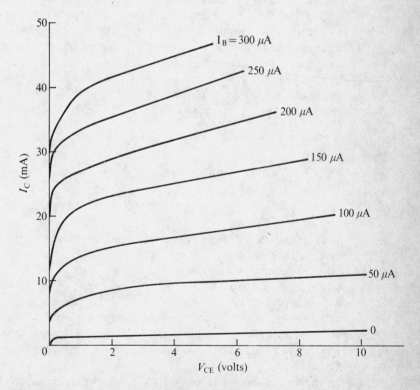

Fig. 2.26. Some common-emitter output characteristics.

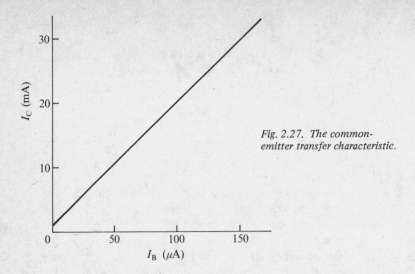

Fig. 2.27. The common-emitter transfer characteristic.

emitter stages are much more nearly matched, and transformer coupling is not required.

The circuit of Fig. 2.25 can also be used to obtain the transfer characteristic, Fig. 2.27. From this characteristic two important parameters may be derived, rather similar to I_{CB0} and α for the common-base circuit. The intercept on the I_C axis is the collector current which would flow with zero base current, sometimes called the common-emitter leakage current. The symbol used is I_{CE0} (pronounced I, C, E, nought) meaning collector current, common-emitter circuit, zero base (input) current. It is considerably larger than I_{CB0} and can be shown to be given by

$$(2.8) \qquad I_{CE0} = \frac{I_{CB0}}{1 - \alpha} .$$

Thus if I_{CB0} is 5 μA and α is 0·995

$$I_{CE0} = \frac{5}{1 - 0·995} = \frac{5}{0·005} = 1000\,\mu A = 1\,mA.$$

In a small transistor operating at a few milliamperes this value of I_{CE0} is quite significant. You may remember that I_{CB0} roughly doubles for a 10 °C rise in temperature, and so, from eqn (2.8), I_{CE0} will also double for this temperature rise. Now a rise in temperature of 10°C might cause I_{CB0} to change from 5 to 10 μA which, if the collector current is a few milliamperes, is not important. But a similar temperature rise in the common-emitter circuit causes I_{CE0} to change, say, from 1 to 2 mA which will be very important. We shall see the real significance of this in Chapter 4.

The second parameter derived from the transfer characteristic of Fig. 2.27 is the current gain. The slope of this line is clearly the ratio of the change in collector current to the change in base current and this is called the small-signal short-circuit forward-current gain and given the symbol h_{fe}. We have referred before to the ratios $I_E : I_C : I_B$ as being $1 : \alpha : (1 - \alpha)$ and obviously the ratio of I_C to I_B (or output to input) is $\alpha/(1 - \alpha)$. This ratio is called h_{FE} or sometimes β (beta). Note that

(2.9)
$$h_{FE} \text{ (or } \beta) = \frac{\alpha}{1 - \alpha}.$$

h_{FE} and h_{fe} are not quite the same. h_{FE} is the ratio of actual d.c. values of collector current and base current whereas h_{fe} is the ratio of changes in these two quantities. In practice there is little difference between h_{FE} and h_{fe}.

It will be found that manufacturers often quote very large ranges for the value of β of a transistor. This is because a small change in α, the parameter which is under the control of the manufacturer, produces a very large change in β. As an example, consider a transistor with an α of 0·995, i.e. a β of $0·995/(1 - 0·995) = 199$. A change in α to 0·996, a 0·1 per cent change (which is very small), produces a change in β to $0·996/(1 - 0·996) = 249$, a change of 25 per cent!

The three parameters h_{fe}, h_{oe}, and h_{ie} are three of the four so-called h-parameters of a transistor. The fourth, h_{re}, is not of great practical significance and will be neglected in this book. The subscripts f, o, and i indicate that h_{fe} is a *f*orward current gain, h_{oe} is an *o*utput conductance, and h_{ie} is an *i*nput resistance, the e in each case indicates that the parameter is associated with the common-emitter circuit.

Thus it can be seen that the common-emitter circuit produces current gain and, because of the fact that the input and output resistances are of the same order, it also produces voltage gain. This will be dealt with in much more detail in Chapter 4.

Common-collector amplifier

It is also possible to use the base as the input but take the output from the emitter, making the collector common to input and output (Fig. 2.28).

This circuit is found to have a very high input resistance (hundreds of kilohms) and a very low output resistance (tens of ohms). It is particularly suitable for the coupling between a source of very high resistance, such as some types of gramophone pick-up, and a low resistance load.

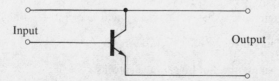

Input

Output

Fig. 2.28. The common-collector configuration.

The common-collector circuit has a voltage gain of just less than one, but a high current gain. The latter must clearly be $1/(1-\alpha)$ as can be seen by considering the current ratios referred to above. A little algebra will show that $1/(1-\alpha)$ is equal to $(1+\beta)$ and hence

(2.10)   Common-collector current gain $= \dfrac{1}{1-\alpha} = (1+\beta)$.

Table 2.1 summarizes the properties of the three circuit configurations. They all satisfy the definition of an amplifier (see Chapter 4) in that they all produce a power gain (which is equal to voltage gain × current gain).

Table 2.1. *Properties of the three circuit configurations*

	Common base	Common emitter	Common collector
Current gain	Less than 1	High	High
Voltage gain	High	High	Less than 1
Power gain	Medium	High	Medium
Input resistance	Low	Medium	High
Output resistance	High	Medium	Low

2.B Test questions

1. The base region of an alloy junction transistor is of thickness of the order of
   (a) $0.5\,\mu m$ (c) $25\,\mu m$
   (b) $25 \times 10^{-12}\,m$ (d) $25\,mm$

2. In the circuit symbol for a p–n junction the arrow always points in the direction of flow of conventional current.
   (a) true
   (b) false

3. In normal operation the junctions of a p–n–p transistor are biased
   (a) both forward
   (b) both reverse
   (c) emitter–base forward, collector–base reverse
   (d) emitter–base reverse, collector–base forward

4. Sketch the diagram of a circuit which could be used to obtain the common-base output characteristics and sketch typical curves.

5. The leakage current in a common base transistor is given the symbol
   (a) I_{CBO} (b) I_{CEO} (c) I_{CCO}

6. The current across the emitter–base junction in a p–n–p transistor consists of
   (a) mainly holes
   (b) mainly electrons
   (c) equal numbers of holes and electrons.

7. Which of the following equations is true when the transistor is operating normally?
   (a) $I_E = I_C + I_B$ (c) $I_C = I_E + I_B$
   (b) $I_B = I_C - I_E$ (d) $I_E = I_B - I_C$

8. A transistor has $\alpha = 0.994$. If $I_C = 2\,mA$ the value of I_E is
   (a) $2.994\,mA$ (c) $2.012\,mA$
   (b) $1.998\,mA$ (d) $1.006\,mA$

9. Sketch a typical common-base input characteristic for a silicon device.

10. Sketch a typical common base transfer characteristic, marking the intercept on the I_C axis.

11. The dynamic output resistance of a common-base transistor could be
   (a) $50\,\Omega$ (c) $10\,M\Omega$
   (b) $26\,m\Omega$ (d) infinite

12. The voltage across the forward-biased emitter–base junction of a germanium transistor is about
   (a) zero   (c) $0.15\,V$
   (b) $0.6\,V$ (d) $10\,V$

13. If the transistor in question 12 is p–n–p the base will be positive with respect to the emitter.
   (a) true   (b) false

14. The common-base amplifier is not a true amplifier because there is no current gain.
   (a) true   (b) false

15. A silicon transistor with $\alpha = 0.99$ is connected as shown in Fig. 2.29. The collector current is approximately
   (a) $2.77\,mA$ (c) $24\,mA$
   (b) $10\,mA$ (d) $10\,\mu A$

16. The nearest value to the base current in the transistor of question 15 is
   (a) $2.8\,mA$ (c) $2.77\,mA$
   (b) $30\,\mu A$ (d) $5.57\,mA$

17. A transistor with $\alpha = 0.995$ is operated in a common-base circuit such that its input resistance is $10\,\Omega$. The collector load resistance is $1\,k\Omega$. The power gain will be approximately
   (a) zero   (c) 99
   (b) just less than one   (d) 10^8

18. Which of the following is most likely to be true?
   h_{oe} of an n–p–n transistor is found to be
   (a) $26\,\mu S$ (c) $10\,M\Omega$
   (b) $26\,m\Omega$ (d) $0.01\,\mu S$

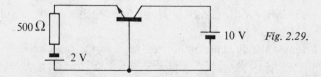

500 Ω   2 V   10 V   *Fig. 2.29.*

19. The best frequency response is obtained by using the transistor in the following configuration
  (a) common base
  (b) common emitter
  (c) common collector

20. Sketch typical output, input, and transfer characteristics for a common-emitter transistor.

21. The relationship between α and β is given by

  (a) $\beta = \dfrac{\alpha}{1 + \alpha}$ (c) $\alpha = \dfrac{\beta}{1 + \beta}$

  (b) $\beta = \alpha(1 - \alpha)$ (d) $\alpha = 1 + \beta$

22. A transistor has a β of 300 but this falls to 150 after a period of time. This represents a fall in α of approximately
  (a) 50 per cent   (c) 40 per cent
  (b) 0·5 per cent   (d) 0·4 per cent

23. A transistor with a β of 199 has a common-base leakage current (I_{CB0}) of $2\,\mu A$. The common-emitter leakage current (I_{CE0}) is
  (a) $400\,\mu A$ (c) about $1\,mA$
  (b) $1{\cdot}99\,\mu A$ (d) less than $1\,\mu A$.

24. The slope of the transfer characteristic of a common-emitter transistor is
  (a) h_{FE} (c) h_{fe}
  (b) α (d) β

25. A transistor with an α of $0{\cdot}995$ is operated in the common emitter configuration such that its input resistance (h_{ie}) is $1\,k\Omega$. The collector load resistance is $2\,k\Omega$. The voltage gain will be approximately
  (a) $0{\cdot}995$ (c) unity
  (b) $3{\cdot}99$ (d) 400

26. The power gain of the amplifier in question 25 is approximately
  (a) 80 000   (c) less than one
  (b) 400   (d) 199

27. A common-collector transistor amplifier has
  (a) high input and output resistance
  (b) high input and low output resistance
  (c) low input and output resistances
  (d) low input and high output resistance.

28. The voltage gain of a common-collector amplifier is
  (a) high
  (b) unity
  (c) less than unity
  (d) about equal to unity.

29. Estimate β for the transistor whose transfer characteristic is shown in Fig. 2.27.

30. Estimate the dynamic output resistance of the common-emitter transistor whose output characteristics are shown in Fig. 2.26 when operating at $V_{CE} = 5\,V$, $I_B = 150\,\mu A$.

2.7 The field-effect transistor

The field-effect transistor (FET) is very different in its mode of operation from the junction transistor considered in the previous section. However, it is simpler in concept.

Basically the current flowing in a bar of semiconductor is controlled by an electric field, rather like the way that the field produced by the grid of a thermionic triode controls the flow of electrons from cathode to anode (see Chapter 9). Indeed, the FET is often referred to as the semiconductor valve, and as such was thought of some years before the junction transistor, although manufacturing technology was such that the junction device was made first.

The main differences between the FET and the junction transistor are:

(1) The FET operates only by the flow of majority carriers and is therefore called a *unipolar* device.
(2) The FET occupies less space in an integrated circuit.
(3) The FET is less noisy. (Noise is discussed in Chapter 4.)
(4) The FET has a very high input resistance.

The main disadvantage is the FET's low gain–bandwidth product. This means that for a given frequency response the gain of the FET will be less than that of a junction transistor. The cost of the FET, which used to be a significant disadvantage, has fallen dramatically in recent years. Modern circuitry frequently includes both FETs and junction transistors, each used where it is the more suitable.

There are two main types of FET, the junction gate FET (JUGFET) and the insulated gate FET (IGFET) more often called the metal oxide semiconductor FET (MOSFET).

The JUGFET

The n-channel JUGFET consists essentially of an n-channel between two heavily doped p-type regions (Fig. 2.30); the ends of the channel being called the *source* and the *drain*, and the p-type material the gate. Conversely, the device could consist of a p-channel with an n-type gate.

The carriers, electrons in an n-channel FET and holes in a p-channel one, flow from source to drain, so that an n-channel FET is operated with the drain positive with respect to the source.

Let us assume that current is flowing due to the voltage V_{DS} between drain and source, the drain being the more positive. Let us further assume that the gate is, for the moment, connected to the source, which is our zero of potential (i.e. $V_{GS} = 0\,V$). There will be a linear drop of voltage from drain to source down the channel (which has a resistance of the order of $1000\,\Omega$). The positive potential of the n-channel with respect to the p-gate reverse biases the gate–channel junction, forming a depletion layer as for any reverse-biased p-n junction. This region, which is almost an

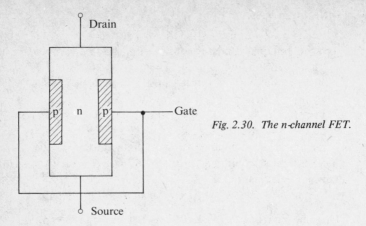

Fig. 2.30. The n-channel FET.

insulator, will be widest at the drain end of the channel, as shown in Fig. 2.31, because the reverse bias will be greatest there.

As V_{DS} is increased the drain current, I_D, will at first increase linearly with voltage, but as it does so the channel gets narrower because of the widening depletion layer (larger reverse bias), and at a certain value of V_{DS} the channel becomes almost shut, or *pinched-off*. It cannot, in fact, quite shut, as the current, which is indirectly producing the depletion layer, would then drop to zero hence opening the channel again. As V_{DS} is increased further the current remains almost constant, just sufficient passing to maintain the channel almost shut.

Thus a graph of I_D against V_{DS} (the output characteristic) has two distinct regions, as will be seen in Fig. 2.32. The first part, on the left, is called the triode region and the horizontal part is called the pentode region.

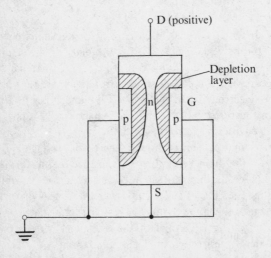

Fig. 2.31. A JUGFET passing current.

These names are because of the similarity of the two regions to the triode and pentode valve characteristics respectively. There is actually a third region where V_{DS} has increased to such a value that, due to the avalanche effect, the drain current rapidly increases and the device might burn out. (This has not been shown.)

If now the gate has a voltage applied which is negative with respect to the source, the depletion layer will be wider for any particular value of V_{DS} because the reverse bias will be greater, and hence I_D will be lower. Characteristics are plotted for different values of V_{GS} and a set of these is shown in Fig. 2.32. If the gate is made sufficiently negative, about $-8\,$V in this example, the channel will be shut completely. The voltage which cuts off the drain current is known as the *pinch-off voltage*.

As can also be seen from Fig. 2.32, if $V_{GS} = 0\,$V then 8 V on the drain will cause the channel to be pinched-off (this is where the characteristic becomes almost horizontal). Similarly at $V_{GS} = -2\,$V about 6 V are needed on the drain to produce pinch-off, etc. A quantity often quoted by the manufacturer is the current flowing at pinch-off when $V_{GS} = 0\,$V, about 3 mA for the device of Fig. 2.32. It is given the symbol I_{DSS}.

The FET acts as an amplifier in that changes in gate voltage produce changes of drain current, which, like the changes of collector current in a junction transistor, pass through a load resistor to give changes in output voltage. This will be discussed in detail in Chapter 4. It should be mentioned here that the input (gate) current is so small that we can often regard it as zero. This is due, of course, to the fact that the gate is reverse biased. The input resistance of a JUGFET is usually more than $100\,\text{M}\Omega$. With such minute input currents it becomes meaningless to talk of current gain!

The output current, I_D, is controlled by the input (gate) voltage, and an

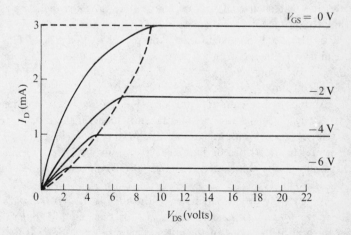

Fig. 2.32. Output characteristics of an n-channel JUGFET.

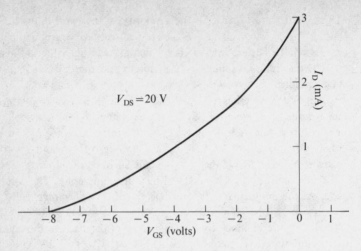

Fig. 2.33. The transfer characteristic of an n-channel JUGFET.

important parameter of the FET is a quantity known as the *mutual conductance* (sometimes called transfer conductance or transconductance) which is the change in drain current divided by the change in gate-source voltage which produces it. This definition assumes that the drain voltage remains constant, which is another way of saying that there is no drain load resistance. Hence

(2.11)   Mutual conductance $= \dfrac{\text{Change in } I_D}{\text{Change in } V_{GS}}$ at constant V_{DS}.

Mutual conductance is given the symbol g_m (g is the normal symbol for conductance). Although the SI unit of conductance is the siemen (S), g_m is often quoted in mA/V, that is the number of mA change in I_D produced by one volt change in V_{GS}. Of course, one mA/V is the same as one mS and it is not incorrect to give g_m in mS.

g_m can be estimated from the output characteristics of Fig. 2.32 by reading off the graph the values of I_D for given values of V_{GS} at a constant V_{DS}. For example, at $V_{DS} = 20\,V$, I_D changes from 3 mA to 1 mA as V_{GS} is changed from 0 V to $-4\,V$, a g_m of 0·5 mA/V.

Alternatively a graph of I_D against V_{GS} (for constant V_{DS}) may be constructed, which is known as the transfer characteristic because it is a plot of input against output. Fig. 2.33 shows such a curve taken from the output characteristics of Fig. 2.32. g_m is obviously given by the slope of this curve. The value of g_m for a JUGFET will probably lie in the range 0·1 to 10 mA/V.

Another important parameter of the FET is its *dynamic output resistance*, r_d (*drain resistance*), which, as for the junction transistor, is the reciprocal of the slope of the output characteristics. For a JUGFET it will usually lie in the range 100 kΩ to 1 MΩ. It is sometimes quoted as a

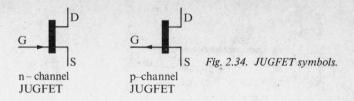

n–channel JUGFET   p–channel JUGFET   *Fig. 2.34. JUGFET symbols.*

conductance, and in American data sheets is often called the *small-signal common-source output admittance* (or conductance), y_{os}. From the $V_{GS} = -2\,V$ curve of Fig. 2.32 I_D falls by about $0.1\,mA$ as V_{DS} falls from $20\,V$ to $10\,V$, an r_d of $(20 - 10)/0.1 = 100\,k\Omega$.

The circuit symbols for both n- and p-channel JUGFETs are given in Fig. 2.34 and it should be noted that the arrow points from the p- to the n-material as for all semiconductor junctions.

Fig. 2.35 gives a circuit which may be used for plotting the output or transfer characteristics. As in Fig.s 2.18 and 2.25 the batteries and potentiometers may be replaced by variable power supplies if this is more convenient.

The MOSFET

There are two main types of metal oxide semiconductor FETs (MOSFETs), the *enhancement* MOSFET and the *depletion* MOSFET. The most commonly available type is the p-channel enhancement MOSFET (n-channel enhancement MOSFETs are harder to manufacture).

Enhancement MOSFET (p-channel)

Two highly doped p-regions are diffused into a lightly doped n-region known as the substrate, or starting material. The p-regions, source, and drain, are separated by about $10\,\mu m$. The gate is a metal plate separated from the substrate by an extremely thin layer of silicon dioxide or other good insulator (Fig. 2.36).

If the gate is made negative with respect to the source a region of holes will be induced below the gate as long as the negative voltage on the gate is above a certain critical value known as the threshold voltage (typically about $-4\,V$). To describe the mechanism by which these holes are induced

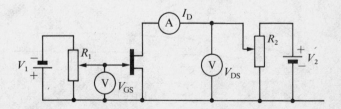

Fig. 2.35.

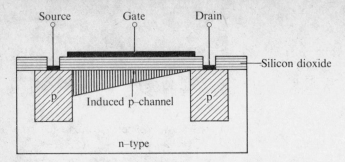

Fig. 2.36. A p-channel enhancement MOSFET.

is beyond the scope of this book, but effectively a p-region, or channel, forms between the source and drain, and with the drain negative with respect to the source, a current of holes flows from source to drain (as in a p-channel JUGFET). As V_{DS} increases negatively the voltage between drain and gate falls (the negative drain voltage getting closer to the more negative gate voltage) and approaches the threshold voltage. At this point the channel becomes pinched-off and the output characteristic produced is similar to that of a JUGFET. The more negative the gate, the wider will be the induced p-channel and the larger the drain current. Output characteristics (I_D/V_{DS}) for an enhancement MOSFET of threshold voltage -4 V are shown in Fig. 2.37. Note that if V_{GS} is, for example, -10 V the device pinches-off at a V_{DS} of -6 V, i.e. a voltage of -4 V between drain and gate (the threshold voltage).

The curves are steeper than for the JUGFET meaning that the value of r_d is lower (between about 1 and 50 kΩ). Mutual conductance values are

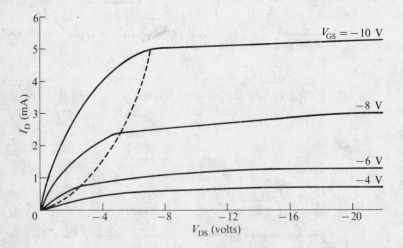

Fig. 2.37. Output characteristics of a p-channel enhancement MOSFET.

Fig. 2.38. *Symbol for p-channel enhancement MOSFET.*

similar (perhaps a little higher) and the input resistance is much higher, because of the insulated gate, and can be as high as $10^{10}\,\Omega$ (ten thousand megohms). The symbol is shown in Fig. 2.38.

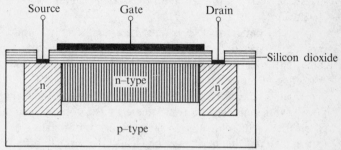

Fig. 2.39. *An n-channel depletion MOSFET.*

The depletion MOSFET

An n-channel depletion MOSFET consists of a p-substrate, two highly doped n-regions (source and drain) and a more lightly doped n-channel between them, as shown in Fig. 2.39. Apart from this n-channel the depletion MOSFET can be seen to be similar to the enhancement type.

At zero V_{GS} current can flow in the channel, being due to electrons flowing towards the drain, which is maintained positive with respect to the source. If the gate is now made negative it induces holes in the n-channel, which makes it less conductive and hence I_D is reduced. It is worth noting that if the gate is made positive with respect to the source more electrons are induced into the n-channel and the drain current increases. The device is then operating in the enhancement mode. Both modes of operation are shown in the output characteristics of Fig. 2.40, and the circuit symbol is shown as Fig. 2.41.

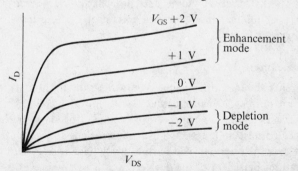

Fig. 2.40. *Output characteristics of an n-channel depletion MOSFET.*

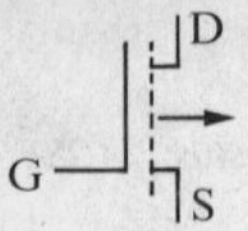

Fig. 2.41. *Symbol for n-channel depletion MOSFET.*

2.8 Practical considerations

Most of the practical aspects of using and handling semiconductor devices should be dealt with in laboratory classes. However a few main points are worth considering here.

All semiconductor devices must be operated only within the temperature range specified, although the storage temperature range may be somewhat wider. The maximum temperature specified will normally be the junction temperature and is in the region of 85 °C for a germanium device and 200 °C for a silicon one. For large devices the manufacturer will often give details of how to calculate the size of any required heat sink. Often part of the device, such as the collector of a transistor, is connected to the case and this must then be electrically insulated from the heat sink or, better, the heat sink insulated from any earthed metal or from other heat sinks.

One time at which heat is likely to enter the device is when soldering the leads into position in the circuit. A thermal shunt is advisable, such as the use of pointed nose pliers between the device and the soldering iron.

Diodes

The manufacturers' data sheets should be consulted to distinguish the anode and cathode. In some diodes the cathode lead is identified by a red spot, this indicating the positive output when being used as a rectifier (see Chapter 3).

Two important quantities which the manufacturer will quote are the peak inverse voltage and the average rectified forward current. The former is the maximum voltage which can be applied in a reverse direction and will be further discussed in Chapter 3. The average rectified forward current speaks for itself. Surge currents will also be quoted, as will such quantities as the maximum forward volts drop and the maximum reverse leakage current.

Zener diodes

For a zener diode various parameters will be quoted. Firstly there is the maximum power dissipation, which must not, of course, be exceeded. Secondly, there is the breakdown voltage together with the slope resistance because, in practice, the characteristic is not quite parallel to the current axis. Also quoted will be the low-voltage leakage current and the temperature coefficient of the device.

Transistors

Most manufacturers will supply a great deal of information on their transistors. They will give a section dealing with absolute maximum values which must under no circumstances be exceeded. These will include the maximum voltages between pairs of electrodes, the maximum collector current and the maximum power dissipation. It should be emphasized, however, that a transistor will not stand the maximum I_C at the maximum V_{CE}. For example a transistor's maximum ratings might include:

Maximum collector-emitter voltage	60 V
Maximum collector current	50 mA
Maximum collector power dissipation	1·2 W

This means that the product of I_C and V_{CE} must not exceed 1·2 W and, clearly, 50 mA at 60 V is 0·05 × 60 = 3 W. Hence, if operated at 60 V, the maximum collector current is 1·2/60 A or 20 mA or if operated at 50 mA the maximum V_{CE} is 1·2/0·05 = 24 V.

The manufacturer will also give the small-signal parameters (such as h_{fe}, h_{ie}, and h_{oe}) and, very important, a means of identifying the three leads. Possibly there will be sets of characteristics and, often, a graph showing how h_{fe} varies with the collector current.

FETs

Remarks made about junction transistors will also apply to FETs. Maximum values will include the maximum gate–source voltage and the parameters will include I_{DSS} and g_m (which may be called y_{fs}).

Great care must be taken in handling MOSFETs. The very thin layer of silicon dioxide may easily be broken down electrically, and even the static charge induced in withdrawing a MOSFET from its package may cause this to happen. Some MOSFETs have a zener diode connected between gate and substrate and often they will have shorting clips between the leads which must not be removed until the device is soldered into position.

2.C Test questions

1. One advantage of the FET over the junction transistor is its very low input resistance.
   (a) true
   (b) false
   (c) it depends on whether it is a MOSFET or a JUGFET

2. A p-channel JUGFET is normally operated with the drain positive with respect to the source.
   (a) true
   (b) false

3. In an n-channel JUGFET the main current flow in the channel consists of
   (a) holes
   (b) electrons
   (c) either, depending on the gate voltage.

4. A JUGFET, in normal operation, has the junction between gate and channel reverse biased.
   (a) true
   (b) false

5. An n-channel JUGFET has a pinch-off voltage of -10 V. If it is operated with $V_{DS} = +8$ V and $V_{GS} = -4$ V will it be operating
   (a) beyond pinch-off (horizontal part of output characteristic)
   (b) before pinch-off (steep part of output characteristic)
   (c) just about at pinch-off.

6. A p-channel JUGFET has a pinch-off voltage of $+8$ V and is operated with $V_{GS} = +5$ V. From the list below pick values of V_{DS} which will cause the device to operate
   (i) before pinch-off
   (ii) beyond pinch-off
   (iii) at pinch-off
   (a) -1 V   (b) -6 V   (c) -3 V   (d) $+6$ V   (e) 0 V
   (f) $+3$ V   (g) $+1$ V.

7. Mutual conductance (g_m) is defined as

   (a) $\dfrac{\text{Change in } V_{GS}}{\text{Change in } I_D}$ (c) $\dfrac{\text{Change in } I_D}{\text{Change in } V_{DS}}$ (e) $\dfrac{\text{Change in } I_G}{\text{Change in } V_{GS}}$

   (b) $\dfrac{\text{Change in } I_D}{\text{Change in } V_{GS}}$ (d) $\dfrac{\text{Change in } I_D}{\text{Change in } I_G}$ (f) $\dfrac{\text{Change in } V_{DS}}{\text{Change in } I_D}$

8. Estimate the value of g_m for the JUGFET whose transfer characteristic is shown in Fig. 2.33, operating at $V_{GS} = -3$ V.

9. Drain resistance, r_d, is defined as:
Use the list (a) to (f) in question 7.

10. The channel is a p-channel enhancement MOSFET is induced into the substrate which consists of
    (a) p-type material
    (b) n-type material
    (c) silicon dioxide.

11. A p-channel enhancement MOSFET with a threshold voltage of -5 V is operated with a value of V_{GS} of -12 V.
If V_{DS} is (i) -3 V, (ii) -15 V, state whether the device is operating on the steep or horizontal part of the output characteristics (i.e. before or beyond pinch-off).

12. Which of the following values of r_d would be reasonable for an enhancement MOSFET
    (a) $0{\cdot}6\ \Omega$
    (b) $100\ \text{M}\Omega$
    (c) $25\ \text{k}\Omega$

(d) too high to measure

(e) too low to measure.

13. One of the main differences between an enhancement MOSFET and a depletion MOSFET is that the depletion device has a source–drain current consisting of minority carriers.

(a) true

(b) false

14. One of the main differences between an enhancement MOSFET and a depletion MOSFET is that in a depletion device the source–drain current flows through a p–n junction.

(a) true

(b) false

15. An n-channel depletion MOSFET is operating with a V_{DS} of +15 V. The gate is +3 V with respect to the source. The device is

(a) operating in the depletion mode

(b) operating in the enhancement mode

(c) cut-off (i.e. zero drain current)

3 Rectification

3.1 Introduction

The domestic electricity supply in this country is 240 V r.m.s., 50 Hz a.c. As you will no doubt be aware from your study of electrical principles there are many reasons for using an a.c. supply. However, there are numerous applications, both in industry and in the home, where we require a d.c. supply.

Three important applications of d.c. in industry are battery charging, electroplating, and traction. The first two clearly require d.c. With regard to traction, the best motor for this purpose is the d.c. series motor, and the locomotives on the 25-kV a.c. railway system are really large, moving rectifier stations!

All these are power applications of rectification, using very large rectifiers which are really outside the scope of this book. Further, they are applications where the d.c. need not necessarily be steady.

The applications of rectifiers which we shall be considering are somewhat different, although employing the same principles. Most electronic equipment operates from a d.c. supply, it usually being essential that in addition to being unidirectional it is a steady or constant supply. It is not very economical to use batteries and their use is normally restricted to portable applications. In this chapter we shall be seeing not only how to produce a d.c. supply but also how to make it almost constant in value. In circuits such as amplifiers, any residual a.c., or ripple as it is called, will be amplified, probably producing serious effects at the output. If a very high degree of stability is required, which will be the case with some circuits, then more sophisticated stabilizers will be needed. These will be considered in Chapter 6 (Feedback).

3.2 Half- and full-wave rectification

The simplest form of rectifier employs a diode connected between the a.c. supply and the load (Fig. 3.1). The semiconductor diode symbol has been used, and will be used in most diagrams in this chapter, it always being understood that a thermionic diode may take its place. In Fig. 3.1 a vacuum diode would require a heater supply which could operate from a

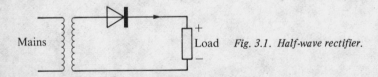

Mains   Load   *Fig. 3.1. Half-wave rectifier.*

suitable secondary winding on the transformer. In the design of vacuum diode circuits, of course, the larger voltage drop across the device will need to be taken into account.

It is convenient to think of the diode as a switch which passes current when its anode is more positive than its cathode, and which is an open circuit at other times. This would be a perfect diode. The very small, but finite, reverse current of a semiconductor diode will not concern us in any of the rectifier applications which we shall be considering, but in practical designs the forward voltage drop, and hence the power dissipated in the device, must be considered.

The voltage waveform across the load in Fig. 3.1 is shown in Fig. 3.2, the peak value, V_p, being almost the same as the peak value of the voltage at the transformer secondary if the diode is a semiconductor. If, as in Fig. 3.1 the load is resistive, the current waveform will be the same shape.

Note in particular the polarity of the voltage across the load. Current flows only in the direction shown, making the top of the load positive with respect to the bottom. Confusion sometimes exists because the positive end of the load is connected to the rectifier's cathode. This is quite logical, in fact, because on the half-cycles when the diode conducts, its anode will be *more* positive than its cathode; hence although the cathode is positive, the anode is even more so. The fact that the cathode is connected to the positive end of the load explains why the cathode is sometimes coded with a red band or spot.

The rectified voltage is, of course, far from constant, but in certain applications this may not matter. It would not, for example, be very important if it were being used to supply a model railway or to charge a battery.

The average value of a half-cycle of a sine wave is twice the peak value divided by π. The waveform of Fig. 3.2 is, however, a sine wave for half the time and zero for the other half and hence its average value will be the peak value divided by π or $0.318\,V_p$.

$$(3.1) \qquad V_{av} = \frac{V_p}{\pi} = 0.318\,V_p.$$

The voltage across the load can be thought of as a d.c. voltage, V_{av}, and, superimposed on it, a waveform of fundamental frequency equal to the supply frequency (usually 50 Hz).

It should be remembered that the voltages of mains transformers are

Fig. 3.2. Half-wave rectifier waveform.

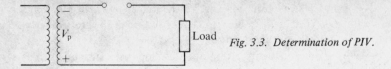

Fig. 3.3. Determination of PIV.

normally quoted in r.m.s. values. Hence if the transformer had a 6·3 V (r.m.s.) secondary (a common value) the peak secondary voltage would be 6·3 $\sqrt{2}$ V or about 8·9 V, and the average value of the rectified voltage, ignoring the drop in the diode, would be 8·9/π or about 2·8 V.

It was emphasized in Chapter 2 that we must not apply a reverse voltage to a diode in excess of that quoted by the manufacturer and we must consider, in any rectifier circuit, what the maximum voltage which will exist across the diode in the reverse direction is during a cycle. This voltage is known as the peak inverse voltage, or PIV, of the circuit.

The value of the PIV can be found by imagining that the diode, on its non-conducting half-cycle, is replaced by an open circuit, as shown in Fig. 3.3. When the diode is not conducting the polarity of the transformer secondary will be as shown and, as no current is flowing and hence no voltage being dropped across the load, this peak voltage will be applied across the diode. Thus the PIV of the half-wave rectifier of Fig. 3.1 is V_p (the peak transformer secondary voltage). This might seem to be an obvious answer, but it is not always the case, as will be seen in some of the later circuits.

Of course, it is possible to omit the transformer and connect the rectifier directly to the mains. In this case the peak voltage would be 240 $\sqrt{2}$ or about 340 V leading to an average load voltage of about 108 V, with a large superimposed a.c. ripple. If this voltage suits the particular application, the cost of the transformer can be saved. However, even in this case, there is an important advantage in using a transformer. This is safety! The mains cannot be directly earthed for various reasons. If a transformer is used, the secondary is completely isolated electrically from the mains and one side can, and usually should, be earthed. Dangers exist when mains transformers are omitted, and you will almost certainly see a notice to this effect on the back of your television set.

Full-wave rectifiers

The half-wave rectifier of Fig. 3.1 only uses the positive half-cycles of the supply. If the diode were reversed it would, of course, be the negative half-cycles and the top of the load would be negative with respect to the bottom. A rectifier circuit designed to use both half-cycles is known as a full-wave rectifier. What happens is that the current flows in *both* half-cycles but the circuit directs it so that it always flows in the same direction in the load.

One type of full-wave circuit, called a *bi-phase* circuit, is shown in

Fig. 3.4. The bottom of the load is shown connected to earth, which will often be the case. It is, in fact, very convenient in any circuit to consider some point as being at zero potential.

The transformer has a secondary which is centre tapped and each half of the secondary has a peak voltage V_p as shown. The transformer manufacturer refers to the voltage of this type of secondary as, for example, 100–0–100 V. This means that the peak voltage *on each half* of the secondary is $100\sqrt{2}$ or $141 \cdot 4$ V.

On one half cycle the polarities will be as shown, diode D1 will conduct but D2 will not. You may find it a help to redraw the circuit with D1 replaced by a short circuit and D2 by an open circuit. In this case the circuit is very much like Fig. 3.1 and, ignoring the diode voltage drop, the peak voltage across the load will be V_p with the polarity shown, current flowing *down* through the load.

On the other half-cycle D2 conducts, but, as will be seen, current still flows *down* through the load, and a sketch of the load voltage against time is given in Fig. 3.5.

The average output voltage is now double that of a half-wave circuit and is given by

(3.2)
$$V_{av} = \frac{2V_p}{\pi} = 0 \cdot 637 \, V_p,$$

where V_p is the peak voltage across *half* of the secondary winding.

However, the ripple superimposed on the average voltage is now at double the supply frequency or, usually, 100 Hz, which makes it easier to reduce as we shall see in §3.3 below.

The PIV of the bi-phase circuit can be found from your drawing showing D1 replaced by a short and D2 by an open circuit. The reverse voltage across D2 will be $2V_p$ because D2 will *see* the two sections of the secondary in series and hence the PIV of this circuit is twice that of the half-wave circuit for the same output voltage.

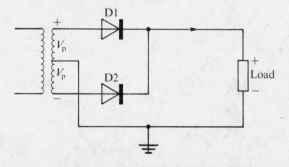

Fig. 3.4. Full-wave rectifier.

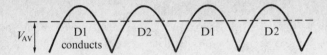

Fig. 3.5. Full-wave rectifier waveform.

The main advantages of this circuit compared with the half-wave one are:

(a)   the ripple, being at twice the frequency, is easier to reduce, or smooth; and

(b)   current flows on successive half cycles in alternate directions in the secondary, which makes the design of the transformer easier.

The main disadvantages are:

(a)   two diodes are required;

(b)   a centre-tapped transformer is more expensive than one with a normal secondary; and

(c)   the PIV is higher.

Fig. 3.6 shows a bridge-rectifier circuit, a method of rectification possessing some advantages over the bi-phase circuit. On the positive half-cycles the current flows as shown, via D1, the load, and D2. On the other half-cycles it flows via D3, the load, and D4. On both half-cycles it will flow *downwards* in the load and hence we have a full-wave rectifier. The PIV may be calculated by drawing D1 and D2 as short circuits and D3 and D4 as open circuits as in Fig. 3.7. The reverse voltage across D3 or D4 is thus seen to be V_p.

The main advantages of the bridge circuit over the bi-phase one are:

(a)   the current in the transformer secondary is a.c.;

(b)   no centre tap is required and a transformer is not absolutely necessary, although you should remember the note on safety above; and

(c)   the PIV is only V_p.

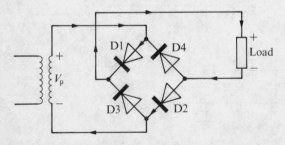

Fig. 3.6. Bridge rectifier.

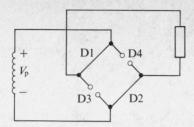

Fig. 3.7. Determination of PIV.

The main disadvantages are:

(a)   there is twice the diode voltage drop; and

(b)   four diodes are used.

Most power supplies for semiconductor circuits employ a bridge rectifier.

All three circuits described perform the function of rectification, although all will require further components to produce the steady voltage needed for electronic circuits. This is discussed in the following section.

3.3 The reservoir capacitor and smoothing

As seen in the previous section, the waveform of the voltage across the load consists of an average value, V_{av}, and a large a.c. component at either the supply frequency in the half-wave circuit or twice it in the full-wave circuit.

The addition of a large capacitor in parallel with the load completely alters this waveform. As we shall see, the capacitor acts rather like a battery or reservoir, charging up via the diode and keeping the load supplied with a current which is much more nearly constant than in the previous circuits.

A half-wave circuit with such a *reservoir capacitor* is shown in Fig. 3.8. This circuit has no load resistor, and it is worth spending a few minutes considering this case before adding the load.

The capacitor will rapidly charge up to the peak of the supply voltage. It actually charges on the positive half-cycles with a time constant of CR, where R is the resistance of the conducting diode and the transformer secondary, but as this resistance will be relatively small the capacitor voltage should reach V_p in a few cycles. After this, as there is a voltage equal to V_p and of the polarity shown, across the capacitor, the diode will

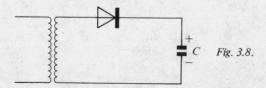

Fig. 3.8.

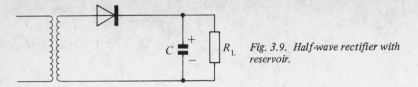

Fig. 3.9. *Half-wave rectifier with reservoir.*

remain non-conducting. Its anode reaches a maximum of V_p, which is just not sufficient to forward-bias it. Hence if the capacitor is perfect and retains its charge, the voltage across it remains at V_p indefinitely and the diode will never conduct after the first few cycles.

If a load is now added in parallel with the capacitor (Fig. 3.9) it discharges through the load resistor R_L with a time constant CR_L which may be relatively large. In this case the capacitor voltage falls below V_p and when the positive-going voltage across the transformer secondary is equal to the falling load voltage diode conduction recommences, recharging the capacitor. The situation, after the initial charging of C, is shown in the waveforms of Fig. 3.10.

The output voltage can be seen to be much more nearly steady than it was, although still not good enough for the power supplies of electronic equipment. It can also be seen that the average output voltage is much higher, being of the order of V_p itself.

It is possible to obtain some idea of the value of V_{av} and the amount of ripple by considering that the shape of the output waveform is a sawtooth as shown in Fig. 3.11. This assumes that the capacitor recharges instantaneously and then discharges linearly, neither of which statements are quite true, of course.

As an example, let V_p be 50 V and the rectifier be operated from the 50-Hz mains. Further, let it be supplying a load current of average value 0·5 A. One cycle occupies 1/50 s or 20 ms. Now the current multiplied by the time for which it flows gives the charge which has left the capacitor. In this case

$$Q = 0\cdot5 \times 20 \times 10^{-3} = 10^{-2}\,\text{C}.$$

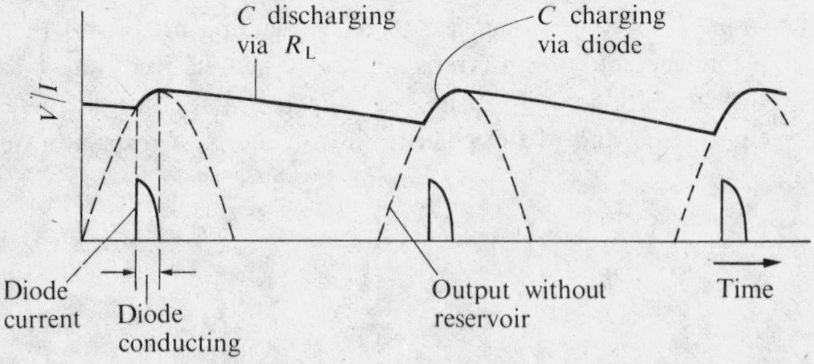

Fig. 3.10. *Voltage across R_L for circuit of Fig. 3.9.*

Hence during each cycle the capacitor loses 10^{-2} C of charge and this loss of charge represents a fall of voltage given by $V = Q/C$. If the value of the capacitor is $2000\,\mu\text{F}$ (a typical value) then the fall in voltage is

$$V = \frac{10^{-2}}{2000 \times 10^{-6}} = 5\,\text{V}.$$

Thus V_A in Fig. 3.11 is $(50 - 5)\text{V} = 45\,\text{V}$ and V_{av} is approximately $47\cdot5\,\text{V}$. The ripple is of peak value $2\cdot5\,\text{V}$, although, as can be seen, it is not sinusoidal. This calculation serves as an interesting example of the application of simple electrical principles.

The reservoir capacitor is a very simple method of vastly improving the ripple, although more improvement is still needed. The ripple is small for small loads and the average voltage is high, being almost equal to V_p. There are, as always, disadvantages.

(a)   The ripple increases with load current and that part of the curve associated with the discharge of the capacitor becomes steeper as the load current increases. In the example quoted, if the load current had been 1 A instead of $0\cdot5$ A the peak ripple would have been 5 V, much greater.

(b)   An important parameter of any power supply is its regulations. This is a measure of how much the output voltage falls with increasing load current owing to the internal resistances. (In a perfect supply the output voltage would not fall at all, of course.) The regulation of a power supply is defined as

(3.3)
$$\frac{V_{\text{no load}} - V_{\text{full load}}}{V_{\text{full load}}} \times 100\%.$$

Obviously the regulation depends on the resistances of the diodes and transformer. However, the reservoir capacitor produces poor regulation for another reason. As the load current increases and the sloping edge of Fig. 3.11 gets steeper V_A and hence V_{av} fall. In the example above, a change of load to 1 A would produce an average output of about 45 V, which is much lower. If 1 A were the full-load current then the regulation would be

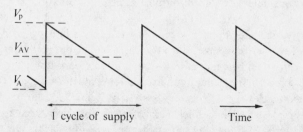

Fig. 3.11.

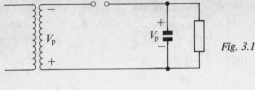

Fig. 3.12. Determination of PIV.

$$\frac{50-45}{45} \times 100 = 11{\cdot}1\%$$

which is not very good.

(c)   The PIV is higher. Fig. 3.12 shows the circuit with the diode removed (non-conducting). The capacitor voltage is almost equal to V_p so that when the supply is at its peak with the polarity shown, the reverse voltage across the diode is $2V_p$. Hence the PIV of this circuit is $2V_p$. Many experimenters have used rectifying circuits without a reservoir capacitor only to find that the diode burns out when a reservoir capacitor is added. You can now see why!

(d)   There is a danger of overheating the diode and the transformer due to the fact that the current through them is in the form of short pulses.

A reservoir capacitor will have a similar effect on a full-wave rectifier, although the ripple now will be much less because the capacitor is re-charged twice as frequently. It has similar advantages and disadvantages although the PIV will not be altered by the addition of the reservoir.

Smoothing

It has been seen that the output of a rectifying circuit with a reservoir capacitor will consist of a d.c. component together with ripple. The d.c. component is the average voltage which will be almost equal to the peak and the ripple will be roughly a sawtooth waveform with a fundamental frequency equal to or double the supply frequency for the half-wave or full-wave circuits respectively.

If we attempted to operate an amplifier from a supply with this amount of ripple, the desired output could be obliterated completely by the 50- or 100-Hz hum, this being an amplified version of the ripple, hence the ripple must be considerably reduced for most electronic applications.

We need to use what is known as a *low-pass filter*. A filter is a circuit which attenuates, or reduces, bands of frequencies in a signal. A low-pass filter passes low frequencies but attenuates higher ones. High and low are relative terms, and in this case we want to pass very low frequencies (in fact d.c. or zero frequency) and attenuate as much as we can the higher ripple frequencies.

The simplest method, which is satisfactory for most purposes, is to put a coil in series with the load to impede, or *choke*, higher frequencies and a capacitor in parallel with the load to short out any higher frequencies

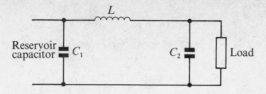

Fig. 3.13. Low-pass filter.

remaining (Fig. 3.13). C_1 is the reservoir capacitor and L and C_2 form the filter.

In the example considered earlier we had a sawtooth wave of amplitude 2·5 V. Although exact calculations are rather complicated, we can gain some idea of the effectiveness of this filter if we consider the ripple to be simply a sine wave of frequency 50 Hz for the half-wave circuit. Of course, the higher frequency components present in the sawtooth wave will be even more attenuated by the filter.

Typical values for this type of supply are 1 H and 2000 μF for L and C_2 respectively. The reactance of 1 H at 50 Hz is

$$2\pi fL = 2.\pi.50.1 = 314\,\Omega.$$

The reactance of 2000 μF is

$$\frac{1}{2\pi fC} = \frac{10^6}{2.\pi.50.2000} = 1\cdot59\,\Omega.$$

This gives a ripple of

$$\frac{1\cdot59}{314-1\cdot59} \times 2\cdot5 = 0\cdot013\,\text{V or 13 mV peak.}$$

The minus sign in this *potential divider* calculation is because the voltages across L and C_2 are 180° out of phase with each other.

Thus the 50 Hz is attenuated whereas the d.c. will not be (if we ignore the resistance of the coil). A similar calculation if the circuit had been a full-wave one would give a ripple of about 3·2 mV peak, a figure which you might like to check.

A circuit diagram of a full-wave rectifier, reservoir capacitor, and low-pass filter is given in Fig. 3.14.

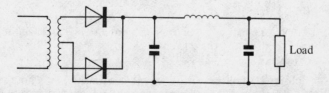

Fig. 3.14. Full-wave rectifier and filter.

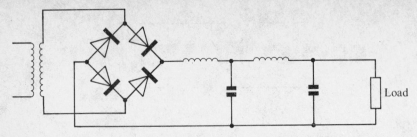

Fig. 3.15. Bridge rectifier with choke input filter.

The capacitors will normally be electrolytic to keep the physical size within reasonable limits. They must be chosen, of course, such that their voltage rating is not exceeded and, as the reservoir capacitor will pass a considerable ripple current, its a.c. current rating must also be considered. Lastly it is essential that the electrolytic capacitors are connected into the circuit with the correct polarity.

Some circuits use a choke-input filter instead of the reservoir capacitor. Such a circuit is shown in Fig. 3.15. At high currents the regulation of this circuit is very good, but it becomes very poor as the current becomes small. The effect of the coil is to try to keep the current constant, but it cannot operate if there are any actual breaks in the current. For this reason the choke-input filter cannot be used with a half-wave circuit. One disadvantage is that the average output voltage is considerably less than the peak voltage and is therefore less than for a circuit employing a reservoir capacitor.

3.A Test questions

1. Draw a circuit diagram of a simple half-wave rectifier feeding a resistive load. Sketch the waveform of the current in the load.

2. A half-wave rectifier, without a reservoir capacitor, is required to produce an average output voltage of 200 V. The r.m.s. voltage of the transformer secondary, ignoring the voltage drop in the diode is
    (a) 444 V   (c) 628 V
    (b) 141·4 V   (d) 63·7 V

3. The output voltage of a half-wave rectifier, with no reservoir capacitor, has a peak value of V_p. The average value is
    (a) V_p/π (c) $V_p/2\pi$
    (b) $2V_p/\pi$ (d) $V_p/\sqrt{2}$

4. The PIV of a half-wave rectifier without a reservoir capacitor is
    (a) V_p/π (c) $2V_p$
    (b) $V_p/\sqrt{2}$ (d) V_p

5. A full-wave, bi-phase rectifier employs a transformer whose secondary is 300–0–300 V r.m.s. If the circuit is feeding a load resistance of 135 kΩ the average output current is

(a) 2 mA   (c) 1·414 mA
(b) 1 mA   (d) 2·828 mA

(Ignore the diode voltage drop and assume that there is no reservoir capacitor.)

6. The PIV of the circuit of the previous problem is
(a) 424 V   (c) 600 V
(b) 848 V   (d) 300 V

7. A full-wave rectifier operating from a 50-Hz supply has a ripple whose fundamental frequency is
(a) 50 Hz
(b) 100 Hz
(c) zero
(d) dependant on the use or otherwise of a reservoir capacitor

8. State three advantages of full-wave, bi-phase rectification.

9. Draw the circuit diagram of a bridge rectifier and give three advantages of this circuit relative to the bi-phase circuit.

10. Explain briefly, the action of a reservoir capacitor when added to a rectifier circuit.

11. Sketch the waveform of the output of a half-wave rectifier with a reservoir capacitor, indicating the times at which the diode is conducting.

12–22 *These questions all deal with a full wave bi-phase rectifier with a reservoir capacitor. It operates from a 50-Hz mains transformer with a 20–0–20 V r.m.s. secondary and is delivering 0·5 A into a resistive load.*

12. Assuming that the ripple is saw-toothed, find the fall in charge per cycle.
(a) 0·005 C   (c) 0·02 C
(b) 5 C   (d) 0·2 C

13. If the reservoir capacitor is 2000 μF the fall in voltage per cycle cycle is
(a) 2·5 V   (c) 1·8 × 10^{-4} V
(b) 0·4 V   (d) 10 V

14. The average output voltage is
(a) 28·3 V   (c) 20 V
(b) 18·75 V   (d) 27 V

15. The ripple will be of peak amplitude
(a) 0·4 V
(b) 2·5 V
(c) 1·25 V
(d) There is not sufficient information given

16. The ripple is sinusoidal.
(a) true
(b) false

17. The fundamental ripple frequency is
(a) 50 Hz
(b) 100 Hz

18. The PIV is
(a) 28·3 V   (c) 20 V
(b) 56·6 V   (d) 14·15 V

19. If the load current were to double the ripple would
(a) double   (c) increase very slightly
(b) halve   (d) not alter

20. If the load current were to double the average output voltage would
(a) double   (c) fall slightly
(b) halve   (d) not alter

21. If the full load current is 1 A the regulation, ignoring the diode voltage drop will be
(a) 8·8 per cent   (c) zero
(b) 100 per cent   (d) 9·7 per cent

22. A low-pass filter consisting of a 1-H coil and a 1000-μF capacitor is now added. Assuming that the ripple is sinusoidal, the peak value of the ripple voltage across the load is
(a) 3·2 mV   (c) 1·24 V
(b) 12·8 mV   (d) 1·24 mV

3.4 Voltage stabilizers

We have seen in the previous sections of this chapter, how to produce a d.c. supply which has very little ripple. However, the output voltage might alter for various reasons.

Firstly, the input a.c. voltage might alter, causing the d.c. output voltage of the power supply to change in proportion. In domestic applications the small changes involved are not very important, at the worst slightly altering the size of picture on a TV screen. However, many industrial applications require that the d.c. supply voltage be stable, and independent of fluctuations of mains voltage.

Secondly, the load current might alter, which, for the reasons outlined in the previous section, will alter the output voltage. Again, this might particularly be the case in industrial situations where various units in a complex piece of equipment are being switched in and out of circuit.

This section will consider simple stabilizing circuits, that is circuits which attempt to maintain the output voltage constant, even though the mains voltage and load current may be varying. As a by-product any ripple which remains will be reduced, because ripple is also a changing output voltage. (In §6.3 we shall be considering rather more advanced, and better, stabilizing circuits.) The circuit that we shall study here employs the constant-voltage characteristic of the reverse-biased zener diode. You will remember that for this device there is a region of the characteristic where the voltage across it is almost independent of the current flowing through it.

Let us assume that the voltage is exactly constant over a certain range of currents, rather than only approximately so. This means that we may pass any current through the device and yet maintain the same voltage across it. It requires a little concentration at first to think of this type of characteristic when we are so used to thinking in terms of components which obey Ohm's Law.

If we connect a perfect reverse-biased zener diode across a load, the load voltage will be held constant. We cannot, however, connect the power supply directly to such a circuit, because if the zener diode had a breakdown voltage of, say, 10 V and we connected it to a nominal 10 V d.c. supply which rose to, say, 11 V, the zener diode would simply pass so much current that it would burn out.

What we have to do is connect a resistor between the supply and the parallel combination of load and zener diode (Fig. 3.16). The circuit operates because the zener diode maintains a constant voltage across the load and itself passes a current, I_z, which will satisfy the other circuit parameters. Actual numbers, rather than letters, will help you to understand what happens.

Let the output voltage, V_o, be 10 V and the load current, I_L, be normally 20 mA. If the normal output voltage of the power supply is 15 V and if we decide to use a value for I_z of 5 mA (the choice will be discussed later) then, because the voltage across the resistor R must be $(15 - 10)$ or 5 V, the value of R is given by

$$R = \frac{5\text{ V}}{(20 + 5)\text{ mA}} = \frac{5\text{ V}}{25\text{ mA}} = 200\ \Omega$$

because the current in it consists of the load current (20 mA) plus the zener current (5 mA). R *does* obey Ohms's law!

We will now consider what will happen if either the power supply voltage or the load current varies. Consider that the power-supply voltage rises to 20 V. As V_o must remain at 10 V, the voltage drop across R must double and become 10 V and this can only occur if the current through it doubles and becomes 50 mA. Now the load current is constant (V_o/R_L) at 20 mA so that the zener current *must* rise to 30 mA. Thus as the power-supply voltage alters, the zener current will alter to keep the voltage drop across R just sufficient to maintain V_o constant. Obviously there are limits to the changes in power-supply voltage which can be stabilized. I_z cannot rise indefinitely or the power rating of the zener diode (given by $I_z \times V_o$)

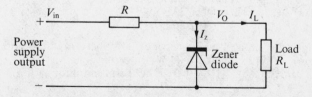

Fig. 3.16. Voltage stabilizer.

will be exceeded. Further it cannot fall below zero. (In practice, to keep the zener diode on its constant-voltage characteristic it cannot fall *quite* to zero.)

As an example, using the figures above, let us calculate the allowable range of power-supply voltages if the maximum zener power is 0·5 W. The *maximum* zener current is $0·5\,W/10\,V = 50\,mA$. With this current the total current in R is $(50 + 20) = 70\,mA$. The voltage drop in $R = 0·07 \times 200 = 14\,V$ and thus the maximum power-supply voltage is $14 + 10 = 24\,V$. The *minimum* zener current is, say, zero and thus the current in $R = 20\,mA$. The voltage drop in $R = 0·02 \times 200 = 4\,V$, leaving the minimum power-supply voltage to be $4 + 10 = 14\,V$. Thus stabilization will take place with a power-supply voltage in the range 14–24 V.

Now consider that the load current alters, due to changes of the load resistance. The load current is nominally 20 mA. Because we are now assuming that the power-supply voltage is constant *and* because the load voltage is also constant, the current in R must remain at 25 mA to drop the required 5 V. If I_L falls to zero, I_z will rise to 25 mA, which is well within its limit of 50 mA calculated earlier. I_L could just about rise to 25 mA, if we assume that I_z could fall to zero. Hence the allowable range of I_L is 0–25 mA.

The circuit can thus be seen to operate, whether the power-supply voltage or load current change, by the zener diode adjusting its current so as to maintain V_o at a constant value.

The calculations above have assumed that the zener voltage is absolutely constant, i.e. that the characteristic is parallel to the current axis. In fact, this is not quite so, and the output voltage will alter slightly, thus entailing the need for the better circuits of §6.3. However, even the simple stabilizer of Fig. 3.16 is very effective.

3.B Test questions

1. The zener stabilizer of Fig. 3.16 is to operate from a power supply of output voltage 25 V. The load current can vary between zero and 30 mA. If the zener voltage is 12 V select the most suitable value for R from the following
   (a) 330 Ω   (c) 470 Ω
   (b) 833 Ω   (d) 390 Ω

2. The minimum zener current in question 1 is
   (a) zero   (c) 33·3 mA
   (b) 3·33 mA   (d) 30 mA

3. The zener diode used in question 1 must have a power rating of
   (a) 0·4 W   (c) 0·04 W
   (b) 0·36 W   (d) there is not sufficient information

4. If the load current of question 1 is zero, calculate the range of input voltage for which the stabilizer will operate if the zener power rating is 0·5 W and the value of R is that found in question 1.

5. Repeat question 4 with a load current of 30 mA.

3.5 Rectifier instruments

Although this chapter has dealt mainly with the rectifier as used in power supplies, a brief mention will be made of its use in measuring instruments.

A very common instrument in an electronics laboratory is the general purpose instrument used to measure a.c. or d.c. currents and voltages and also to measure resistance. The *heart* of such an instrument is a moving-coil meter, which measures d.c. or the average value of a.c. Obviously it cannot measure a sinusoidal current directly because it would measure the *true* average of a complete cycle which is zero. Hence the current to be measured needs to be full-wave rectified by a bridge circuit (Fig. 3.17), and the meter reads the average value.

Voltage is, of course, measured by measuring the current produced in a known resistor.

One important word of warning! Most commercial instruments, although *measuring* the average value, are *calibrated* in r.m.s. values on the assumption that the current being measured is sinusoidal. This is done because we are often more interested in the r.m.s. (or power-producing) value. However the calibration will only hold as long as we are measuring sinusoidal quantities. In electronics ·we often deal with non-sinusoidal currents and voltages and great care must be taken in interpreting the meter readings in this case.

The relationship between r.m.s. and average values of a sine wave is given by

(3.3) $I_{rms} = 1 \cdot 11\, I_{av}.$

Thus the readings have, in effect, been scaled up by a factor of 1·11. If we take a reading of a non-sinusoidal current or voltage all that we can say with certainty is that the *average* value is found by dividing the reading by 1·11 (this giving the full-wave rectified average).

As an example, assume that an instrument of such a type measures 10 V when connected to a non-sinusoidal wave. The average value is 10/1·11 or 9 V.

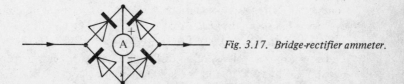

Fig. 3.17. Bridge-rectifier ammeter.

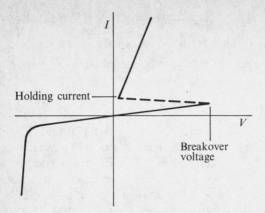

Fig. 3.18. Thyristor characteristic.

If we know the actual wave-shape we may, of course, be able to calculate the true r.m.s. value. For example a waveform we meet frequently in electronics is a square wave. This is a unique waveform in that the r.m.s., average, and peak values are all equal. Hence if our instrument were reading 10 V we would know that because the average value was 9 V the r.m.s. value would also be 9 V.

3.6 Controlled rectification

Whilst it is not the intention in this book to deal at length with what is often called high-power electronics, no elementary consideration of electronics would be complete without some mention of the *silicon-controlled rectifier* (SCR). It is a device which can be used to control the flow of current to a load, not always a simple task when dealing with heavy currents!

The SCR (thyristor)

The SCR is a four layer device, p–n–p–n, the outer layers being the anode (p) and cathode (n). If the anode is made negative with respect to the cathode the two outer junctions share the applied voltage and only a small reverse current will flow, unless the applied voltage is so high that it breaks down these junctions. With the anode positive with respect to the cathode, most of the applied voltage appears across the reverse-biased centre junction, and it is found that at a certain voltage the device becomes conducting as long as the current is maintained above a minimum value called the *holding current*, I_h. The voltage for this break down is called the *breakover voltage*. A typical thrystor characteristic is drawn in Fig. 3.18.

The breakover voltage of an SCR will usually be a few hundred volts, possibly up to 1000 V, and the holding current will range from less than a

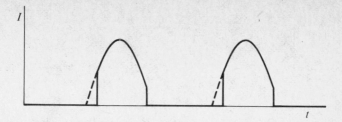

Fig. 3.19.

milliampere for a small device to perhaps 50 mA for a large one. The voltage across the *conducting* thyristor will be of the order of 0·7–1·25 V depending on the current.

If an SCR is used in a half-wave rectifier circuit with an input of peak value *less than* the breakover voltage, it would clearly never conduct. However, if the peak value is *greater than* the breakover voltage it will start to conduct when the applied voltage exceeds this value and will stay conducting until the current falls below the holding current. The output current waveform in this case is shown in Fig. 3.19.

An SCR has a third connection, known as a trigger or gate, which may be made to the inner p- or n-region. If connected to the inner p-region (next to the cathode) it is called a *cathode gate* and if to the n-region an *anode gate*. Cathode gate SCRs are the more common.

A current introduced at the gate will, if in the right direction, *lower* the breakover voltage and thus we can control the point on the input cycle voltage at which conduction starts. The characteristics of a cathode-gate SCR are shown in Fig. 3.20.

The gate current required to trigger the device need only flow for a few microseconds and pulses are usually used for the purpose. Positive pulses

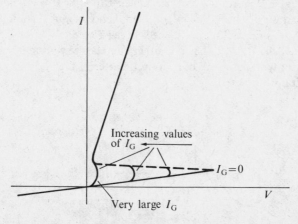

Fig. 3.20. Effect of gate current.

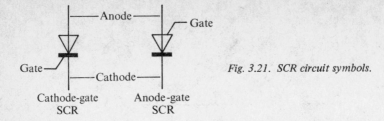

Fig. 3.21. SCR circuit symbols.

are needed for a cathode-gate device, negative for an anode-gate one. The circuit symbols of both types are given in Fig. 3.21. A simple circuit using an SCR to control the flow of current to a load is shown in Fig. 3.22.

The control circuit will generate pulses to trigger the SCR at a particular point on the cycle, thus controlling the power in the load. A typical application is the control of power to a lighting circuit and hence the brightness of the lamps in the circuit. Whilst this could be done with a variable resistor, the power wasted in the resistor may be considerable, presenting enormous problems, and cost, if used to control a large lighting installation such as in a theatre. As the SCR is, in effect, a switch, it does not waste power when *open*.

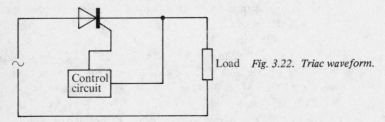

Fig. 3.22. Triac waveform.

With the circuit of Fig. 3.22 we can still only supply current for half the cycle at the most, and a device similar to the SCR is available, which will breakover in both directions, and hence supply a current when triggered, as in Fig. 3.23. This device is called a *triac*.

It should be realized that the rather complex waveforms produced by either the thyristor or the triac do not matter in the type of application for which they are used.

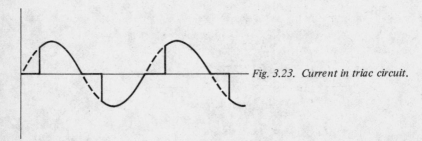

Fig. 3.23. Current in triac circuit.

3.C Test questions

1. A moving-coil meter measures
   (a) r.m.s. values
   (b) average values
   (c) average values multiplied by 1·11
   (d) r.m.s. values divided by 1·11

2. The average value of a sine wave of peak value V_p is
   (a) $V_p/\sqrt{2}$ (c) $2V_p/\pi$
   (b) zero   (d) $V_p/1\cdot11$

3. A voltage-measuring instrument consists of a perfect diode, a 100 kΩ resistor, and a moving-coil meter of negligible resistance in series. If it is connected to an a.c. supply of peak value 100 V, the meter will read
   (a) 1 mA   (c) 0·707 mA
   (b) 0·636 mA   (d) 0·318 mA

4. Why can we not use a moving-coil meter in series with a diode to measure an alternating current?

5. Sketch the circuit diagram of a bridge-rectifier ammeter using a moving coil meter.

6. A bridge-rectifier ammeter uses four diodes of negligible resistance and a moving-coil meter also of negligible resistance. The reading on the meter when connected in series with a load carrying a sinusoidal current of 10 A r.m.s. will be
   (a) 7·07 A   (c) 9·0 A
   (b) 14·14 A   (d) 10·0 A

7. A rectifier instrument calibrated to read the r.m.s. value of a sine wave, is connected to a non-sinusoidal voltage of r.m.s. value 100 V. It will read
   (a) 100 V
   (b) 141·4 V
   (c) 111 V
   (d) there is not sufficient information

8. If the waveform of question 7 has a form factor of 1·01 the reading given will be
   (a) 100 V
   (b) 109·9 V
   (c) 99 V
   (d) there is not sufficient information

9. A voltage below breakover is applied to an SCR. State across which junctions the applied voltage appears with the anode
   (a) positive
   (b) negative with respect to the cathode.

10. An SCR has its gate electrode connected to the inner p-region. Is it a cathode- or anode-gate device?

11. When a pulse is used to trigger a cathode-gate SCR does the pulse of current (conventional) flow into or out of the gate?

12. Sketch the circuit diagram of an SCR controlling the power in a resistive load.

13. In the circuit of question 12, what are the minimum and maximum portions of the applied sinusoidal waveform that can appear across the load?

4 Amplifiers I

4.1 Introduction

There are various types of amplifier and they are used for many purposes. We shall talk of small-signal amplifiers and large-signal amplifiers, low-frequency amplifiers and high-frequency amplifiers, voltage amplifiers and current amplifiers. One property, however, which they all have is the ability to amplify power, that is to produce *power gain*.

The term 'power gain' tends to imply that we increase the power or, in effect, that we get something for nothing. This is obviously not true, and an amplifier might better be described as a controller of power. It takes a signal at one power level and produces an identical signal, in theory, at a higher power level. An amplifier forces a d.c. source to supply a varying power to a load, varying in a way determined by the input signal.

Imagine, for example, that we connect a battery, a variable resistor, and a loudspeaker in series. Now in theory we could twist the control of the variable resistor in such a way that the current in the circuit varied sinusoidally at, say, 100 times a second, or 100 Hz. The sinusoidal variation would be superimposed on a steady direct current which might, in fact, damage the loudspeaker, but we are only talking in theoretical terms. Obviously we could not move the spindle of the potentiometer at 100 Hz by hand, but if we could we would hear a 100-Hz note from the loudspeaker. We have controlled a d.c. source in such a way that it is producing power at 100 Hz. In theory it is possible to rotate the potentiometer spindle in such a way as to produce any sounds we wish to hear, even, say, a complete symphony by Beethoven!

Our amplifier simply does this for us, acting as a variable resistor controlled by a very small voltage or current.

This chapter deals firstly with amplifiers in general and then goes on to consider, in some detail, single-stage amplifiers using transistors and FETs. Chapter 5 will deal with multistage amplifiers, radio frequency (r.f.) and power amplifiers, and finish with a brief consideration of integrated-circuit amplifiers.

4.2 Properties of amplifiers

Before we consider actual circuits it will be as well to get some idea of the properties of an amplifier, and in later sections we shall see how different devices behave in relation to these properties.

Gain

This is the main consideration for most people when thinking of amplifiers.

Gain is the change in the output quantity divided by the change in the input quantity which produces it. The quantity referred to can be voltage, current, or power, leading to the concept of voltage gain, A_v; current gain, A_i; and power gain, A_p. (*A* stands for amplification.) Power gain is the product of voltage gain and current gain, hence

(4.1)
$$A_p = A_v A_i.$$

The voltage step up which may be produced by a transformer must not be confused with gain. If the transformer steps *up* voltage it will step *down* current by, at least, the same amount, giving a power gain of unity (if 100 per cent efficient). It cannot be overemphasized that although we may use amplifiers as voltage or current amplifiers they all ultimately amplify power.

To take, as an example, a well-known system, consider a record player using a crystal cartridge which might be developing a voltage of, say, 0·5 V for a particular signal on a disc. The input resistance of the amplifier will be about 2 MΩ, a value specified by the manufacturer of the cartridge. Hence

$$\text{Input current} = \frac{0.5 \text{ V}}{2 \text{ M}\Omega} = 0.25 \, \mu\text{A};$$

$$\text{Input power} = 0.25 \, \mu\text{A} \times 0.5 \text{ V} = 0.125 \, \mu\text{W}.$$

The output signal required might be 1 W into a speaker of resistance 4 Ω (a standard value).

$$\text{Required power gain} = \frac{1}{0.125 \times 10^{-6}} = 8\,000\,000.$$

Now $P = \dfrac{V^2}{R}$;

hence the output voltage V is given by

$$V = \sqrt{(PR)} = \sqrt{(1 \times 4)} = 2 \text{ V},$$

and the voltage gain $= \dfrac{2}{0.5} = 4.$

The output current I will be given by

$$I = \frac{V}{R} = \frac{2 \text{ V}}{4 \, \Omega} = 0.5 \text{ A},$$

and the current gain $= \dfrac{0.5}{0.25 \times 10^{-6}} = 2\,000\,000.$

Note that $2\,000\,000 \times 4 = 8\,000\,000$, the power gain. Power gain is usually expressed in decibels:

$$\text{Power gain in decibels} = 10 \log_{10} \frac{P_{\text{out}}}{P_{\text{in}}} \quad \text{dB}.$$

In our example,

$$\text{Power gain} = 10 \log_{10} (8 \times 10^6)$$

$$= 10 \times 6 \cdot 9$$

$$= 69 \text{ dB}.$$

Input and output resistances

Many students, particularly those new to electronics, tend to ignore the input and output resistances of an amplifier, yet these parameters can be just as important as the gain.

Input resistance is the resistance which would be measured across the input terminals of the amplifier. It will, in fact, actually be an impedance because of capacitative effects, but we will assume that the frequencies with which we are concerned are low enough for capacitive effects to be ignored.

Consider again the record player with its crystal cartridge. It might produce 1 V and have an internal resistance of $2 \text{ M}\Omega$ (this is the resistance of the material of which the cartridge is made). If the amplifier has an input resistance of $2 \text{ M}\Omega$, as in the example above, the voltage at the amplifier's input terminals will be $0 \cdot 5$ V because of the potential divider action of the two $2\text{-M}\Omega$ resistances.

However, if an amplifier with an input resistance of, say, 50Ω had been used, a value typical for the common-base circuit referred to in Chapter 2, the situation shown in Fig. 4.1 would exist, and

$$V_{\text{in}} = 1 \text{ V} \times \frac{50}{2\,000\,050} \simeq 25 \,\mu\text{V}.$$

This is a quite ridiculous situation, attenuating 1 V to $25 \,\mu\text{V}$ before amplification.

Clearly, the input resistance of 50Ω is much too low for a source of this internal resistance. Similarly, if the signal which we wanted to amplify

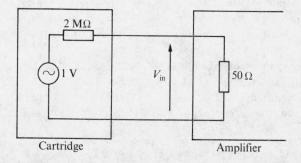

Fig. 4.1.

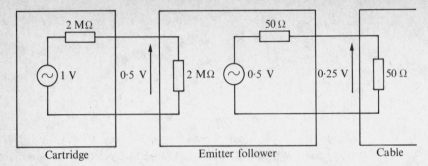

Fig. 4.2.

had been a current, the 50-Ω input resistance might have been too high. Thus input resistance is a parameter which has to suit the requirements.

An amplifier also has an output resistance and we can conveniently consider the output as being a voltage generator in series with a resistor, the output resistance. An amplifier developing, say, 2 V with an output resistance of 1 MΩ and feeding a 4-Ω loudspeaker would develop only

$$\frac{4 \times 2}{1\,000\,004} \text{ V} = 8\,\mu\text{V}$$

across the load, a situation just as ridiculous as that of Fig. 4.1.

An amplifier used frequently in electronics is a circuit known as an emitter or source follower. You will meet this circuit a great deal, but all that needs pointing out here is that it is a circuit with a very high input resistance and a very low output resistance. It also has a voltage gain of slightly less than one. A typical example of its use is when we wish to feed the output of a high-resistance source, such as a cartridge, into a low-resistance load. The latter might be a long length of cable preceding the main part of the amplifier. For reasons which are rather complex and difficult to understand, a length of cable has a low input impedance, such as 50 Ω.

We have seen above that if we connected a 2 MΩ cartridge directly to 50 Ω the input voltage would be 25 μV. However, if we insert an emitter follower with an input resistance of 2 MΩ and an output resistance of 50 Ω between the cartridge and the cable we get the situation shown in Fig. 4.2.

The voltage gain of the emitter follower is assumed to be one, and the various voltage levels will be as shown, resulting in a signal of 0·25 V into the cable. Although the voltage gain of the whole system is 0·25 (1 V generated by the cartridge and 0·25 V into the cable), there is a power gain:

$$\text{Cartridge output power} = \frac{1^2}{4 \times 10^6} = 0·25\,\mu\text{W}$$

$$\text{Power into cable} = \frac{(0 \cdot 25)^2}{50} = 1 \cdot 25 \text{ mW}$$

which gives a power gain of 5000 or 37 dB.

These examples should convince you that not only are input and output resistances important, but that they are intimately connected with the power gain.

Distortion

The input and output signals of an amplifier should be identical in every respect except in amplitude. Any alteration in the shape of the signal is *distortion*. We shall not consider distortion in detail, but will mention two different causes of it.

First, frequency distortion, or the fact that all frequencies are not amplified by the same amount (see Chapter 5). This would not be directly observable for a sinusoidal input. However, any other recurring waveform consists of the fundamental and a number of harmonics (Fourier), so that if all the frequencies present are amplified by different amounts the shape of the waveform will change, perhaps drastically. Also different frequencies will be shifted in phase by differing amounts.

Another cause of distortion is the fact that the characteristics of the device will not, in practice, consist of equally spaced straight lines as we often assume. This produces non-linear distortion and leads to various effects. First, signals of different amplitudes are amplified by different amounts, so that if we double the input we shall not double the output; secondly, harmonics of the input frequencies are produced; and lastly frequencies equal to the sums and differences of the various input frequencies occur. The latter two forms of distortion are called *harmonic* and *intermodulation distortion*, respectively.

Usually the non-linear distortion produced by an amplifier is measured by applying a sinusoidal input and expressing the sum of the harmonics produced as a percentage of the fundamental. This is called the total harmonic distortion (THD) and to give you some idea of the figures, the THD of a well-known high-fidelity amplifier on the market is 0·03 per cent at 1 kHz at full output power.

Noise

Noise may be defined as any unwanted signal. Hence the most beautiful music is noise if it is not the signal output that we require. It does not, of course, need to be audible noise, so that the white spots on a television picture are also called noise.

Noise can be external or internal. External noise comes from a number of sources, some man-made, some not. Examples of man-made noise are:

(a)   the 50-Hz hum picked up from power leads;

(b)   random noises, such as the 'clicks' produced when other apparatus is switched off or on; and

(c)   ignition interference from cars, etc.

Non-man-made noise can come from many sources, lightning being a typical example.

However, even if all sources of external noise could be eliminated, there would still be noise produced inside the amplifier itself. You have probably all discovered that even if an amplifier has no input, then, with the gain turned up, he output will consist of a general hiss, often described as being like fat frying in a pan.

Every resistor in the circuit contains electrons which carry thermal energy and noise is produced because there are slight variations in the amount of this energy carried by each electron. This is called *Johnson* or *thermal noise* and the noise power produced is proportional to the absolute temperature.

Shot noise exists because the direct currents flowing in the circuit are not absolutely constant, but fluctuate very slightly due to the fact that the current is carried in discrete lumps (called, of course, electrons). There is also the so-called *flicker noise*, a source of noise not fully understood and which produces a noise power inversely proportional to frequency.

The level of noise in a circuit is only important relative to the level of the signal. One mV of noise would not matter too much if the signal were 10 V but would be disastrous with a signal of 1 μV!

The amount of noise relative to the signal is expressed as the *signal-to-noise ratio*. This expresses, in decibels, the ratio of signal power to noise power. Thus the specification of an amplifier might include a signal-to-noise ratio of 60 dB at its full-power output of 30 W. 60 dB is a power ratio of 10^6 : 1 so that the noise output will be 30 W/10^6 or 30 μW.

Noise sets a limit to the amount of amplification which is possible and makes the amplification of very small signals difficult or even impossible. It is a very important factor in electronics and telecommunications. Whilst a detailed study of noise is beyond the scope of this book, its effects must never be ignored.

4.A Test questions

1. Define gain, as applied to an amplifier.

2. An amplifier will always produce
   (a) power gain   (c) voltage gain
   (b) current gain   (d) all three

3. Power, voltage, and current gains are related by
   (a) $A_v = A_i \times A_p$ (c) $A_i = A_v \times A_p$
   (b) $A_p = A_v \times A_i$ (d) $A_p = A_v + A_i$

4. Explain why a transformer with a voltage step up of 100 cannot be regarded as an amplifier with a voltage gain of 100.

5–12

An accelerometer, a device which gives an output proportional to the acceleration of the mechanism to which it is attached, develops an output voltage of 2 mV and has an internal resistance of 1·5 MΩ. It is connected via an amplifier to a moving-coil meter of resistance 0·5 Ω and full-scale deflection of 10 mA. The amplifier has an input resistance of 500 kΩ and an output resistance of 0·5 Ω. These questions involve the various gains, and should show you that the gains must be clearly defined.

5. The voltage across the input terminals of the amplifier is
    (a) 2 mV   (c) $\frac{1}{3}$ mV
    (b) 0·5 mV   (d) 1 mV

6. To produce full-scale deflection in the meter the voltage across the output terminals of the amplifier is
    (a) 5 mV   (c) 20 mV
    (b) 10 mV   (d) 2·5 mV

7. To produce this voltage, the voltage of the imaginary generator in series with the amplifier's output resistance must be
    (a) 10 mV   (c) 6·67 mV
    (b) 5 mV   (d) 2·5 mV

8. The voltage gain of the amplifier when connected to the meter, i.e. the voltage across the output terminals divided by the voltage across the input terminals, is
    (a) 2·5   (c) 15
    (b) 3·33   (d) 10

9. The current gain of the system is
    (a) 10^6 (c) 10^{-6}
    (b) 10   (d) 10^7

10. The power gain of the amplifier, meaning the power in the load divided by the power into the input terminals, is
    (a) 80 dB   (c) 74 dB
    (b) 87 dB   (d) 93 dB

11. The voltage gain of the *system*, i.e. from the 2 mV of the accelerometer to the voltage across the load is
    (a) 2·5   (c) 15
    (b) 7·5   (d) 5

12. The power gain of the *system*, i.e. the power in the load in relation to the power produced by the accelerometer is
    (a) 80 dB   (c) 74 dB
    (b) 87 dB   (d) 93 dB

13. Define distortion.

14. Explain what is meant by frequency distortion, and state whether it would affect the shape of
    (a) a sine wave
    (b) a square wave

15. Discuss the cause of, and the effects produced by, non-linear distortion.

16. Define noise.

17. Discuss some of the causes of external noise.

18. Johnson noise is the noise produced by
    (a) lightning
    (b) the fact that current is carried by discrete particles
    (c) the fact that electrons in a resistor carry slightly different amounts of energy

19. Shot noise is the noise produced by
[Use the same key as in question 18.]

20. Johnson noise gives a noise power
    (a) proportional to the absolute temperature
    (b) inversely proportional to the absolute temperature
    (c) independent of temperature
    (d) proportional to the square of the absolute temperature

21. An amplifier is developing a signal voltage of 10 V across a load. If the noise voltage is 1 mV, the signal-to-noise ratio is
    (a) 40 dB   (c) 80 dB
    (b) 100 dB   (d) −80 dB

22. An amplifier is producing a noise power of 100 μW.
If the signal-to-noise ratio is 40 dB, the output power is
    (a) 1 W   (c) 2 W
    (b) 4 mW   (d) 0·5 W

23. An amplifier has a signal-to-noise ratio of 50 dB.
If the output power is 10 W into a 15 Ω load, the noise *voltage* across the load is
    (a) 39 μV   (c) 0·12 mV
    (b) 1·5 mV (d) 39 mV

4.3 The small-signal common-emitter amplifier

Amplifiers are divided into many types. One such division is into small-signal and large-signal amplifiers, terms which can be defined in various ways. The important point is that small-signal amplifiers are ones where the signal is such that the device is operating on the linear part of its characteristics. Hence the parameters which have been defined in Chapter 2 will apply and distortion will, in theory, be zero.

There are two distinct parts in the design of an amplifier, or in the analysis of an amplifier circuit. First, we must consider the d.c. operation, that is we must ensure that the device is operating correctly in a d.c. sense, and that the d.c. voltages and currents are of suitable magnitude and polarity. This must be done with reference to the various characteristics. Secondly we can consider the effect which the amplifier produces on the a.c. input signal. For this we can ignore the d.c. conditions as long as we are able to assume that the device is still operating on the linear part of its characteristics. We can then replace the practical circuit with a small-signal equivalent circuit.

We have seen in Chapter 2 how a transistor may be used in three different configurations, common emitter, common base, and common collector (or emitter follower). By far the most frequently used is the common emitter, because it provides both voltage *and* current gain and because it has input and output resistances that for most applications are neither too high nor too low. A discussion of the other two configurations will be given later in this chapter.

Firstly consider an n-p-n transistor in series with a load resistor connected to a d.c. supply as shown in Fig. 4.3. If a p-n-p transistor were used the supply V would be negative with respect to earth and the collector current, I_C, would flow in the opposite direction. The base–emitter connections are open circuited. Clearly, by Kirchhoff's second law

(4.2)

$$V_{CE} = V - I_C R_L,$$

and a load line representing this equation can be drawn on the output characteristics (V_{CE} against I_C) as shown in Fig. 4.4.

To make it clearer, numbers have been used in this diagram, assuming that $V = 10\,V$ and that $R_L = 2\cdot5\,k\Omega$. The intercepts on the axes can easily be found using eqn (4.2):

$$\text{if } I_C = 0, \ V_{CE} = 10\,V;$$
$$\text{if } V_{CE} = 0, \ I_C = V/R_L = 10\,V/2\cdot5\,k\Omega = 4\,mA.$$

The value of R_L might be fixed by other parts of the circuit, but if it is not then a value can be determined such that it will give a suitable load line as in Fig. 4.4, bearing in mind the limiting parameters of the transistor.

This load line now enables us to find the values of I_C and V_{CE} for any given value of I_B. Thus if I_B is $5\,\mu A$, I_C will be about $0\cdot95\,mA$ or if I_B is $20\,\mu A$, I_C will be about $3\cdot7\,mA$. In these cases V_{CE} will be about $7\cdot6\,V$ and $0\cdot8\,V$ respectively.

It is worth noting at this point that the higher collector current ($3\cdot7\,mA$) is flowing at the lower collector voltage ($0\cdot8\,V$). This must occur because at the higher current more voltage is dropped across R_L leaving less across the transistor. It means that the resistance of the transistor is

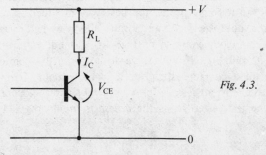

Fig. 4.3.

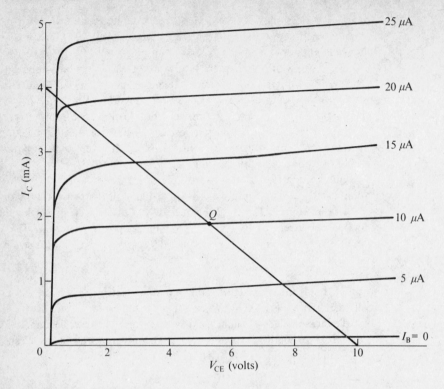

Fig. 4.4.

being reduced by the flow of base current, which illustrates the way in which the base current controls the resistance of the transistor. It was explained in Chapter 2 that a junction transistor amplifies because the base–emitter junction is forward biased and the base–collector junction reverse biased. In the n–p–n transistor of Fig. 4.3, a forward-biased base-emitter junction means that the base is about 0·6 or 0·7 V positive with respect to the emitter, assuming a silicon transistor, or in other words, that current flows, in a conventional sense, into the base terminal.

Now we require this base current to vary, because it is the variations of base current which are our input and which cause the changes in collector current. As most of the waveforms which we shall wish to amplify have equal positive and negative peaks, the best way in which to operate the transistor is to choose the bias point about halfway along the useable part of the load line. In our example of Fig. 4.4 a standing base current of 10 μA would be suitable. Of course, we do not have to operate the transistor exactly on a characteristic, but it is convenient to do so as an example. This d.c. current is called a *bias current* and the operating point on the load line is referred to as the *quiescent point*, marked Q on the characteristics.

The obvious way to obtain this bias current is from the d.c. supply, V, which is already there, and a resistor R_1 connected from the supply to the base will, if of the correct value, supply the required current. This is shown in Fig. 4.5.

We can easily estimate the value of R_1 if we remember that the base voltage in this case will be about 0·6 or 0·7 V positive. This means that the voltage across R_1 is $(V - 0·6)$ and if the bias required is I_B the value of R_1 is given by

(4.3)
$$R_1 = \frac{V - 0·6}{I_B} .$$

In our example

$$R_1 = \frac{(10 - 0·6)\,V}{10\,\mu A} = 0·94\,M\Omega$$
$$\text{or } 940\,k\Omega.$$

This beautifully simple method of bias has very serious disadvantages as we shall see later. However, for the moment let us assume that we are using the circuit of Fig. 4.5.

We are now in a position to consider the a.c. operation, that is the action of the correctly biased transistor as an amplifier. It is convenient to consider the input as being sinusoidal, although in many applications the waveforms we wish to amplify will be much more complex. However, a sine-wave analysis of the circuit enables us to predict its effect on these more complicated waveforms.

The transistor is a current-operated device, and the collector current is proportional to the base *current* rather than the base *voltage*. We will therefore assume that the input to the base consists of a sinusoidal current, and it is usual to feed this current via a capacitor which is large enough to act as a short circuit at the frequencies which are to be amplified. The capacitor will block the d.c. on the base from the signal source, which it could possibly damage. Also any d.c. fed from the signal source may well modify the quiescent conditions. The output characteristics are redrawn in Fig. 4.6 and again the bias point Q is shown.

Obviously the base current can fall to zero without reverse biasing the

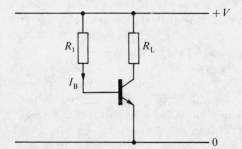

Fig. 4.5. Simple bias.

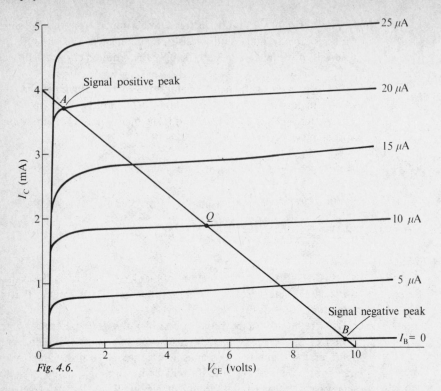

Fig. 4.6.

base–emitter junction, that is a fall of $10\,\mu A$. Because Q is near the centre of the load line the base current can also rise by $10\,\mu A$, so that the input can be of peak value $10\,\mu A$ or peak-to-peak value $20\,\mu A$. This is, however, the maximum, because if the base current fell below zero and become negative, or if it rose above $20\,\mu A$, the output waveform would be clipped, or severely distorted. Equally obviously an input of this magnitude would not have been possible with Q further up or down the line, because although one half-cycle would then have been amplified satisfactorily the other would have been clipped.

Thus with the maximum possible input signal the transistor operates between the points A and B on the load line. The collector current is varying from $0.1\,mA$ to $3.7\,mA$, a peak-to-peak swing of $3.6\,mA$ and this therefore represents a current gain of $3.6\,mA/20\,\mu A = 180$. That is $A_i = 180$. It is worth noting here that at this bias point the ratio of the d.c. values of I_C to I_B is about 200. This, of course, is β and the current gain (a.c.) can be seen to be slightly less than β.

The output (collector) voltage swings from $0.8\,V$ to $9.8\,V$, a peak-to-peak output of $9.0\,V$. Note that the largest voltage occurs for the smallest collector current, a fact mentioned earlier. This means that as the input goes *more* positive the output becomes *less* positive representing a $180°$ difference of phase between output and input voltage.

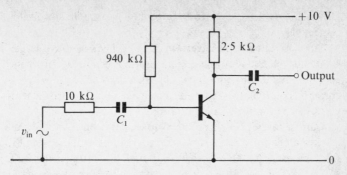

Fig. 4.7. Common-emitter amplifier.

The voltage gain is rather harder to calculate because, as has been said, the transistor is a current-operated device. If the signal source is a voltage, as will often be the case, it must be fed to the base via a resistor which is large compared to the input resistance of the transistor. If this is not done, the non-linear input characteristic (Fig. 2.24) will distort the input current and hence the output will be distorted. As the input resistance of a small transistor in the common-emitter configuration will be in the order of $1\,k\Omega$ a resistor of about $10\,k\Omega$ would be suitable, as shown in Fig. 4.7.

The input peak-to-peak current is $20\,\mu A$ and so the input voltage v_{in} is approximately $(20\,\mu A \times 11\,k\Omega) = 0.22\,V$, peak to peak, and the voltage gain, A_v, is $9.0\,V/0.22\,V = 41$. In using the value $11\,k\Omega$ it has been assumed that the input resistance of the actual transistor is $1\,k\Omega$.

Although, in electronics, we usually find it more convenient to work in peak or peak-to-peak values, sometimes root-mean-square (r.m.s.) values are required. For a sine wave the r.m.s. value is the peak divided by $\sqrt{2}$ or the peak-to-peak divided by $2\sqrt{2}$.

In the example considered above the peak-to-peak output-signal current was $3.6\,mA$ and this has an r.m.s. value of $3.6/(2\sqrt{2}) = 1.27\,mA$. Similarly the r.m.s. value of the output-signal voltage is $9/(2\sqrt{2}) = 3.18\,V$ and the output-signal power is thus $3.18 \times 1.27 = 4.04\,mW$.

The source is developing a peak-to-peak current of $20\,\mu A$, which is $20/(2\sqrt{2}) = 7.07\,\mu A$ r.m.s. into a total resistance of $11\,k\Omega$ representing an input power of I^2R or $(7.07)^2 \times 10^{-12} \cdot 11 \cdot 10^3 = 550 \cdot 10^{-9}\,W$. Thus the power gain is

$$A_p = \frac{4 \cdot 10^{-3}}{550 \cdot 10^{-9}} = 7273 \text{ or } 39\,dB$$

Note that $A_i \times A_v = 180.41$ which is also 39 dB.

It is worth remembering that the values of voltage and power gain have been calculated from the source. If one considers the actual base terminal of the transistor as the input, the r.m.s. voltage input will be $7.07\,\mu A \times 1\,k\Omega = 7.07\,mV$ and the power input becomes $7.07\,mV \times 7.07\,\mu A =$

50×10^{-9} W. This leads to voltage and power gains of 449·8 and 49 dB respectively.

This last calculation is very misleading because the voltage at the base is not sinusoidal due to the non-linear input characteristic of the transistor, and the conversion from peak-to-peak values to r.m.s. values is no longer correct. You will, nevertheless, often see it performed in textbooks, and you should be aware of its limitations.

Quite a considerable amount of space has been devoted to the common-emitter amplifier because it is a circuit which is fundamental to an understanding of amplifiers. We shall find in section 4.6 that the FET behaves in a similar fashion, although it is, of course, a voltage-operated device.

4.B Test questions

1. Define the term 'small-signal amplifier'.

2. The transistor whose output characteristics are shown in Fig. 2.26 is operated from a 10-V supply and has a collector load resistor of 250 Ω. Copy the characteristics and draw a load line.

3. Choose a suitable bias current if the input is a sine wave of peak value 100 μA. Give reasons for your choice.

4. Determine the quiescent collector current and voltage.

5. Assuming that the transistor is silicon, which of the following would be the most suitable value for R_1 if the circuit of Fig. 4.5 were being used?
   - (a) 10 kΩ
   - (b) 100 kΩ
   - (c) 94 kΩ
   - (d) 940 kΩ

6. Which value of R_1 given in question 5 would be the most suitable if the transistor were germanium?

7. If the input is 100 μA peak, the r.m.s. output signal current is
   - (a) 25·5 mA
   - (b) 18 mA
   - (c) 9 mA
   - (d) 70·7 μA

8. The current gain is
   - (a) 255
   - (b) 127·5
   - (c) 175
   - (d) 361

9. The r.m.s. output signal voltage is
   - (a) 3 V
   - (b) 6·4 V
   - (c) 4·5 V
   - (d) 2·3 V

10. The output signal power is
   - (a) 82 mW
   - (b) 163 mW
   - (c) 0·23 mW
   - (d) 20·7 mW

11. The input resistance (h_{ie}) of this transistor is 500 Ω. If the signal is fed to the base via a 4-kΩ resistor as in the circuit of Fig. 4.7, estimate the voltage gain.
   - (a) 5·1
   - (b) 7
   - (c) 70
   - (d) 14·2

12. With the same circuit as in question 11 estimate the power gain in dB.
   (a) 0·01 dB   (c) 65·1 dB
   (b) 7 dB   (d) 29·6 dB

13. Make a rough estimate of the power gain of the actual transistor from the base) and explain why this estimate is not very reliable.
   (a) 39 dB   (c) 7 dB
   (b) 29·6 dB   (d) 65·1 dB

14. Estimate the average power being supplied by the 10-V supply.
   (a) 0·175 W
   (b) 20·7 mW
   (c) 0·1 W
   (d) there is not sufficient information.

15. Estimate the average power being dissipated in the transistor using the key of question 14.

4.4 Biasing the common-emitter amplifier

It was mentioned earlier that the method of bias used in section 4.3 is not at all satisfactory. In this section we shall consider why this is so and what can be done to overcome the problem.

The circuit of Fig. 4.5 produces a base bias current which only depends on the supply voltage and the resistor R_1. In other words, as long as these values remain constant the bias current will not alter and the circuit is for this reason called a *constant base-current* circuit. This might, at first, seem to be just what is required. Unfortunately two things can happen which may upset the situation, perhaps disastrously.

The circuit should be designed to operate with any transistor of the type being used; this being clearly a property of any well-designed circuit. It was seen in Chapter 2 that the value of β can vary greatly for transistors of the same type, perhaps by a factor as large as 3:1.

Let us see what would happen if the transistor in the circuit of Fig. 4.5 were replaced by one with a β of 400 instead of 200. The characteristics of Fig. 4.4 would now become something like those of Fig. 4.8. However, the load line is *not* a function of the transistor, but only depends on V and R_L and hence will not alter. Neither will the bias current, which will remain at 10 μA. Q will now be almost at the top of the load line and the maximum possible input current will be negligible. The circuit will now simply not work as an amplifier. Of course, if the value of β were to fall, the operating point could be near the bottom of the load line, with similar consequences.

The second effect to be considered is that of temperature. Quite apart from any change in ambient temperature, the transistor will heat up as current flows through it, and, as has been explained in some detail in §2.6, this will cause the leakage current to rise, roughly doubling for every 10 °C rise in temperature. The effect is that the characteristics all expand

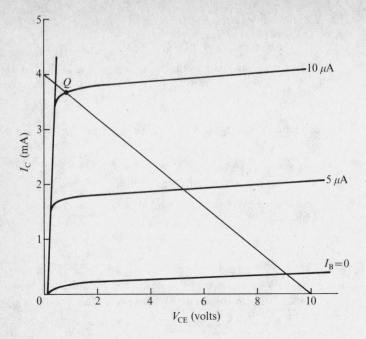

Fig. 4.8.

upwards producing a move in the position of Q similar to that produced by the rise in β just discussed, and with the same results. It is even possible to get an unstable situation where a rise in temperature produces a rise in collector current which in turn produces a further rise in temperature and so on leading to the transistor burning out. This is called *thermal runaway*.

What is required is a bias circuit which keeps the *collector current* constant, and hence the position of Q on the load line remains constant. One such *constant collector-current* circuit is shown in Fig. 4.9.

The circuit operates as follows, numbers being used to make the argument easier to follow. The potential divider R_1 and R_2 keeps the base at a voltage, with respect to earth, given by

$$\frac{R_2 V}{R_1 + R_2} = \frac{16 \times 10}{84 + 16} = 1 \cdot 6 \, \text{V}.$$

This calculation has ignored the base current, but as the current in R_1 and R_2 is about $100 \, \mu\text{A}$, the $10 \, \mu\text{A}$ of base current is not very significant. Now, if the base voltage is $1 \cdot 6 \, \text{V}$, the emitter voltage is about $0 \cdot 6 \, \text{V}$ less, at $1 \cdot 0 \, \text{V}$ (assuming a silicon transistor). Hence the current in R_E must be $2 \, \text{mA}$ ($1 \, \text{V}$ across $0 \cdot 5 \, \text{k}\Omega$) and the collector current will also be about $2 \, \text{mA}$. This has been calculated without any knowledge of the value of β for the transistor. The actual base current will depend on β, so that if $\beta = 200$, $I_\text{B} = 2 \, \text{mA}/200 = 10 \, \mu\text{A}$, but if β rises to 400, I_B will fall to $5 \, \mu\text{A}$.

The circuit will oppose any change in I_C, thus if I_C tends to rise *for any reason*, the emitter voltage will rise, *reducing* the base-emitter voltage and hence reducing the base current. This fall in base current reduces the collector current which therefore tries to stay constant.

We can thus see that I_C is fixed by the circuit values and I_B depends on β instead of in the simpler circuit, where I_B is fixed by the circuit values and I_C depends on β.

The circuit will work better the lower the values of R_1 and R_2, because then the base current flowing via R_1 is even less significant. However we must be careful because if we lower R_1 and R_2 too much they will tend to short out the incoming signal and a compromise must be made.

One last point; this bias circuit tries to keep the collector current constant whatever is causing it to change. But the whole point of the amplifier is to produce a changing collector current. What this means is that the stabilizing circuit considerably reduces the gain of the amplifier. It is, in fact, an example of negative feedback, a subject to be discussed in Chapter 6. Now, the stabilizing effect which we require is a d.c. effect, and if we are only concerned with the amplification of a.c. we can eliminate it from an a.c. point of view by shorting out R_E with a large capacitor. Sometimes, however, this is not done, and the loss of gain is the price paid to obtain the advantages of negative feedback discussed in Chapter 6.

An estimate of the capacitance required is often obtained by ensuring that the reactance of the capacitor is, at most, one-tenth of R_E at the lowest frequency to be amplified. Thus if the amplifier of Fig. 4.9 is used down to 40 Hz, say, the reactance of the capacitor would need to be at the most 50 Ω at 40 Hz. Thus

$$C_E = \frac{1}{2 \cdot \pi \cdot 40 \cdot 50} \text{ F} = 80 \,\mu\text{F}$$

and we would probably use $100 \,\mu\text{F}$, a standard value. The complete amplifier is shown in Fig. 4.10.

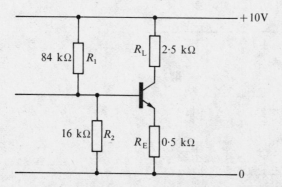

Fig. 4.9. Emitter bias.

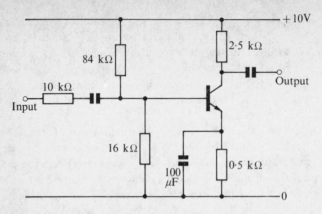

Fig. 4.10. Common-emitter amplifier.

4.C Test questions

1. Discuss the disadvantages of the method of bias used in Fig. 4.5.

2. A transistor is biased with a constant base current of $10\,\mu A$, and the value of α at the operating point is 0.992. The collector current will be
(a) $9.92\,\mu A$ (c) $0.08\,\mu A$
(b) $1.24\,mA$ (d) $4\,mA$

3. If the transistor of question 2 is changed for one with $\alpha = 0.995$, estimate the new value of I_C.
(a) $2\,mA$ (c) $1.24\,mA$
(b) $9.95\,mA$ (d) $0.05\,\mu A$

Questions 4–10 refer to Fig. 4.11.

4. Estimate the value of the base voltage, with respect to earth, assuming a silicon transistor with a value of V_{BE} of $0.6\,V$.
(a) $2\,V$ (c) $1.4\,V$
(b) $2.6\,V$ (d) $0.6\,V$

5. Find the value of R_2 if the current in it is $100\,\mu A$.
(a) $1.4\,k\Omega$ (c) $26\,M\Omega$
(b) $14\,k\Omega$ (d) $26\,k\Omega$

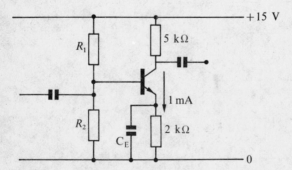

Fig. 4.11. Questions 4–10 (4.C).

6. If $\beta = 100$, the current in R_1 is
   (a) 100 μA (c) 90 μA
   (b) 110 μA (d) 10 μA

7. The required value of R_1 is
   (a) 136 kΩ (c) 138 kΩ
   (b) 112·7 kΩ (d) 100 kΩ

8. The quiescent value of V_{CE} is
   (a) 8 V (c) 15 V
   (b) 10 V (d) 9·4 V

9. If the amplifier is designed to operate over a range of frequencies from 100 Hz to 100 kHz, a suitable value for C_E would be
   (a) 1 μF (c) 8 μF
   (b) 0·1 μF (d) 0·008 μF

10. If C_E is omitted the effect would be
    (a) a rise in the voltage gain
    (b) a fall in the voltage gain
    (c) a great increase in distortion
    (d) a less stable circuit

11. The circuit of Fig. 4.12 shows a transistor with the base bias current supplied from the collector voltage. If the required bias is 50 μA, the quiescent collector current is 1 mA and R_L is 5 kΩ estimate a suitable value of R_B assuming a silicon transistor.
    (a) 88 kΩ (c) 100 kΩ
    (b) 188 kΩ (d) 1 MΩ

12. Considering the circuit shown in the previous question, try to work out the train of events which would follow a rise in collector current caused by, say, an increase in temperature. Consider what happens to the collector voltage, base current, and hence the collector current. Would this circuit tend to stabilize the operating point of the transistor?

4.5 Common-base and common-collector circuits

Although the transistor is most frequently used with the emitter common to input and output, it can be used with either the base or the collector common.

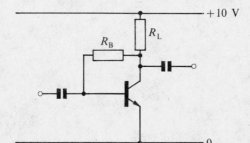

Fig. 4.12. Questions 11–12 (4.C).

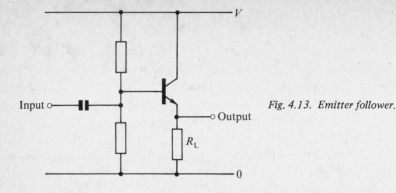

Fig. 4.13. Emitter follower.

Common base

This is probably the least-used configuration. As was seen in Chapter 2 the current gain of the amplifier is less than unity, its input resistance is low and its output resistance is high. While it can be used to match a low-resistance source to a high-resistance load, its main use is in certain very high-frequency applications because its high-frequency response is better than that of the other two configurations.

Miller discovered that the high-frequency performance of an amplifier is seriously affected by the capacitance between input and output. The base of a transistor is between the emitter and the collector and, if the latter are the input and output respectively and the base is common or earthed, it greatly reduces this capacitance and improves the high-frequency performance.

Common collector

This configuration is often called the *emitter follower*, because the output (emitter) is in phase with, or *follows*, the input. An emitter follower is shown in Fig. 4.13.

It can be seen from the circuit that the voltage gain is about unity, because if the base–emitter voltage were *exactly constant* at, say, 0·6 V, the base voltage *changes* would appear directly as emitter voltage *changes* with a resultant voltage gain of one.

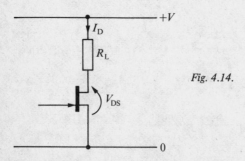

Fig. 4.14.

Reference to Table 2.1 (p.28) shows that this configuration has a high input resistance and low output resistance, and it is therefore very frequently used in the type of situation described in detail in section 4.2 and shown in Fig. 4.2.

4.6 The FET amplifier

The operation of the FET as an amplifier can be dealt with in a very similar fashion as for the junction transistor, the main difference being that the current through it (drain current) is controlled by a voltage rather than a current.

Fig. 4.14 shows an n-channel JUGFET in series with a load resistor. As for the transistor, a load line may be drawn on the output characteristics, and this has been done in Fig. 4.15 for a load of 2 kΩ and a supply voltage of 20 V. If the gate is biased at −2 V the quiescent point Q will be approximately midway between V and $V_{GS} = 0$, the standing drain current being 4 mA and the standing value of V_{DS} being 12 V.

For the moment it will be assumed that the −2-V bias on the gate is

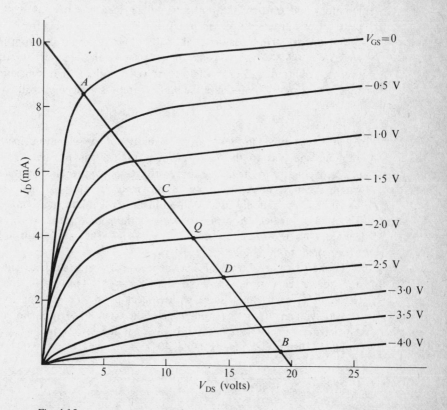

Fig. 4.15.

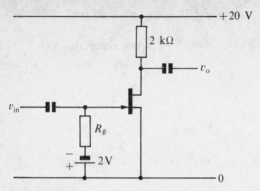

Fig. 4.16.

achieved by using a battery, as shown in Fig. 4.16, the resistor R_g being required to stop the battery shorting out the incoming signal. As the gate current will be extremely small, R_g will not drop any appreciable voltage.

We can obviously now have an input signal of peak value 2 V (a peak signal voltage of more than 2 V would forward bias the gate–channel junction) and thus we may operate on the load line from A to B. This peak-to-peak input of 4 V will produce a drain current swing from 0·5 mA to 8·4 mA or 7·9 mA peak-to-peak, and a drain-voltage swing from 19·1 V to 3·3 V or 15·8 V peak-to-peak, out of phase with the input. The voltage gain A_v produced is 15·8/4 = 3·95. In the case of the FET it is meaningless to talk of current gain because the input current is negligibly small.

Biasing

A battery, as shown in Fig. 4.16, is obviously not a very convenient way of applying the bias to the gate. Further, the voltage required is of the *opposite* polarity to the drain supply voltage, unlike the junction transistor where the base needs a supply of the *same* polarity as the collector.

The most commonly used method of bias is that known as *source bias*. The gate is earthed via a resistor R_g and the source lifted to a positive voltage by a resistor R_s as shown in Fig. 4.17. The gate is thus negative *with respect to* the source.

In the example above, the bias needed is 2 V and the quiescent-drain and source current is 4 mA, so that $R_s = 2\,\text{V}/4\,\text{mA} = 0·5\,\text{k}\Omega$. The capacitor C_s earths the source with respect to the a.c. to stop the bias varying with an alternating input signal. Like C_E in Fig. 4.10 it should have a reactance of, at most, one tenth of R_s at the lowest frequency to be amplified. If this were, for example, 40 Hz,

$$C_s = \frac{1}{2 \cdot \pi \cdot 40 \cdot 50}\,\text{F} = 80\,\mu\text{F}$$

and again we would probably use $100\,\mu\text{F}$.

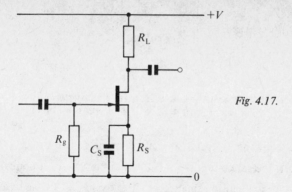

Fig. 4.17.

Sometimes the gate is raised to a slightly positive voltage, the source being raised to a higher value to maintain the correct bias. This is done to stabilize the bias point against changes in the characteristics from one device to the next, and the method is shown in Fig. 4.18.

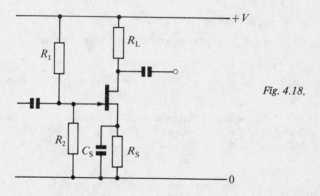

Fig. 4.18.

The method shown in Fig. 4.17 cannot be used for an enhancement mode MOSFET, because in this case a bias of opposite polarity is needed. The method of Fig. 4.18 can be used, as long as the gate is made positive with respect to the source for an n-channel device. Alternatively, the method shown in Fig. 4.19 can be used.

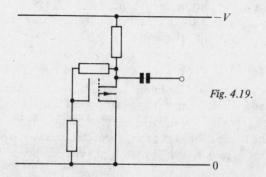

Fig. 4.19.

The FET can also be used in the common gate, and common drain (source follower) configurations.

4.D Test questions

Questions 1–11 concern a FET whose output characteristics were shown in Fig. 2.32. It is in series with a 7·5-kΩ load and is connected to a 20-V supply.

1. Copy the characteristics and construct a load line.

2. If the gate bias is to be −4 V find the quiescent point.

3. If the input is a sine wave of peak value 2 V estimate the swing in drain current.

4. Find the r.m.s. value of the drain signal current.

5. Find the peak-to-peak value of the drain voltage.

6. Find the r.m.s. value of the signal-output voltage.

7. Estimate the voltage gain.

8. Estimate the signal power developed in the load.

9. Find the average power developed by the 20-V supply.

10. Find the value of the source resistor required to provide the bias.

11. If the method of bias shown in Fig. 4.18 is to be used and if R_1 is 19 MΩ and R_2 is 1 MΩ find the new value of the source resistor to maintain the bias at −4 V.

4.7 Equivalent circuits

Equivalent circuits, or models, are used throughout electrical engineering. The idea is to represent a complex device by a circuit consisting of simple elements, although it is not normally possible to construct the simple circuit, except in theory.

A simple example is the representation of a *real* battery by a *perfect* battery in series with a resistor which represents the internal resistance of the real battery. This is a trivial example, but one we often use without even thinking of it as an equivalent circuit.

Transistor

As long as it is operating on the linear part of its characteristics, a transistor can be thought of as a current generator whose current is controlled by the input current. The input, or base, current is flowing into the input resistance of the device. Thus the common-emitter transistor can be represented, to a first approximation, by the equivalent circuit of Fig. 4.20.

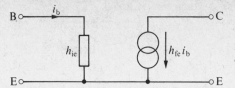

Fig. 4.20. Simple equivalent circuit of transistor.

The input resistance is h_{ie} and the current gain is h_{fe} (see Chapter 2). This circuit only deals with the a.c. operation of the transistor and gives no information regarding the d.c. conditions.

Figure 4.10 may be redrawn, replacing the transistor by the equivalent circuit and omitting any components that have no effect on the a.c. operation. The result is shown in Fig. 4.21.

C_1, C_2, and C_3 are omitted because they are short circuits at the frequencies at which the circuit operates. C_3, of course, shorts out R_E which is therefore also omitted. R_1 and R_2 are large compared with h_{ie} and can often be omitted, as they have been in this case. Lastly, the d.c. supply voltage V is omitted since the equivalent circuit is essentially an a.c. circuit. The power supply feeding the circuit will have a large capacitor across it (see Chapter 3) which will act as a short to a.c.

If we assume that h_{ie} is $1\,k\Omega$ then i_b will be $v_{in}/11\,k\Omega$. Now v_o is produced by the current of the generator, $h_{fe}\,i_b$, flowing upwards in the load. It makes the top end of the load instantaneously negative with respect to ground when the input voltage is instantaneously positive with respect to ground. Clearly these two voltages are $180°$ out of phase. Thus

$$v_o = h_{fe} \cdot i_b \cdot R_L$$

and, if h_{fe} is 200,

$$v_o = 200 \cdot \frac{v_{in}}{11\,k\Omega} \cdot 2 \cdot 5\,k\Omega$$

$$= 45 \cdot 5\,v_{in}.$$

Hence

$$A_v = \frac{v_o}{v_{in}} = 45 \cdot 5, \text{ with } 180° \text{ phase shift.}$$

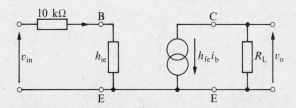

Fig. 4.21. Equivalent circuit of Fig. 4.10.

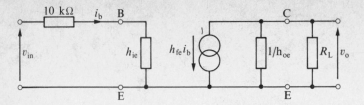

Fig. 4.22.

This is a little higher than we calculated in §4.3 because we have ignored the fact that the transistor has an output resistance. The output resistance of the particular transistor we are considering is about $23\,\text{k}\Omega$, a value of h_{oe}, the output conductance, of about $43\cdot5\,\mu\text{S}$. This resistance is in parallel with the generator, as shown in Fig. 4.22.

The resistance R_L in parallel with h_{oe} gives an equivalent load of

$$R_{eq} = \frac{23 \times 2\cdot5}{23 + 2\cdot5} = 2\cdot25\,\text{k}\Omega, \text{ giving}$$

$$A_v = \frac{200}{11} \times 2\cdot25 = 41,$$

which is the value estimated from the characteristics. As long as we are sure that the transistor is operating linearly, the use of the equivalent circuit for estimating the gain is much easier than drawing, and using, the output characteristics.

A rather more advanced use of the equivalent circuit is in estimating the gain and input resistance of an emitter follower. Making the same assumptions as for the common-emitter circuit, and ignoring h_{oe}, the equivalent circuit of Fig. 4.13 is shown in Fig. 4.23. Note that the emitter is now connected to earth via R_L.

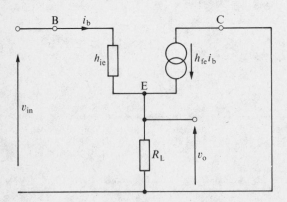

Fig. 4.23. Equivalent circuit of emitter follower.

The current in R_L is $(i_b + h_{fe} i_b)$ or $i_b(1 + h_{fe})$. Applying Kirchhoff's second law to the input circuit gives

$$v_{in} = i_b h_{ie} + R_L i_b (1 + h_{fe}),$$

and the input resistance is given by

(4.4)

$$R_{IN} = \frac{v_{in}}{i_b} = h_{ie} + R_L(1 + h_{fe}).$$

With typical values, $h_{ie} = 1\,k\Omega$, $h_{fe} = 200$, and $R_L = 2\cdot5\,k\Omega$

$$R_{IN} = 1 + (2\cdot5 \times 201)$$

$$= 503\cdot5\,k\Omega$$

very high for a junction transistor.

Now,

$$v_o = i_b(1 + h_{fe})R_L$$

thus

$$A_v = \frac{v_o}{v_{in}} = \frac{i_b(1 + h_{fe})R_L}{i_b h_{ie} + R_L(1 + h_{fe})i_b}$$

(4.5)

$$= \frac{R_L(1 + h_{fe})}{h_{ie} + (1 + h_{fe})R_L}.$$

With the figures quoted

$$A_v = \frac{2\cdot5 \times 201}{1 + (201 \times 2\cdot5)} = 0\cdot998.$$

There is, of course, a current gain. It will be $(h_{fe} + 1)$ or in this case 201, giving a power gain of $201 \times 0\cdot998 = 200\cdot6$ or 23 dB.

Field-effect transistor

The equivalent circuit for the FET will be similar to that for the junction transistor, but the input resistance is so high that it can be regarded as infinite. The current generator is this time controlled by the gate–source voltage and, from eqn (2.11), its value will be $g_m v_{gs}$. Thus we arrive at the circuit of Fig. 4.24.

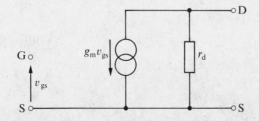

Fig. 4.24. *Equivalent circuit of FET.*

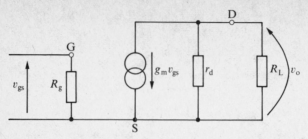

Fig. 4.25. Equivalent circuit of Fig. 4.17.

The amplifier circuit shown in Fig. 4.17 has the equivalent circuit of Fig. 4.25. v_o is produced by the current $g_m v_{gs}$ flowing in the parallel combination of r_d and R_L. If we call this combination R then

$$v_o = g_m \cdot v_{gs} \cdot R$$

and

$$A_v = \frac{v_o}{v_{gs}} = g_m R, \text{ with } 180° \text{ phase shift.}$$

Let us take R_L as $2\,k\Omega$ as in the example in section 4.6 and let us assume that r_d is so high relative to $2\,k\Omega$ that we can ignore it. From the characteristics of Fig. 4.15, g_m is seen to be of the order of $2.5\,mA/V$ (at Q). Hence $A_v = 2.5 \times 2 = 5$.

This is somewhat higher than we calculated earlier because in operating from A to B in Fig. 4.15 we are to some extent operating on the non-linear part of the characteristics. If we had restricted the input to $1\,V$ peak-to-peak and operated between C and D the gain would have been much closer to 5, a fact which you might like to check.

4.E Test questions

1. A transistor has $h_{fe} = 120$ and $h_{ie} = 1.5\,k\Omega$. The input signal is fed to the base via a 4.5-$k\Omega$ resistor. If h_{oe} is negligible, find the load resistance required to produce an overall voltage gain of 100.

    (a) $5\,k\Omega$ (c) $80\,k\Omega$
    (b) $100\,k\Omega$ (d) $1.5\,k\Omega$

2. If, in question 1, h_{oe} cannot be ignored but is equal to $100\,\mu S$, find the new value of load resistance to keep the gain at 100.

    (a) $5\,k\Omega$ (c) $50\,k\Omega$
    (b) $10\,k\Omega$ (d) $3\,k\Omega$

3. With the load resistance of question 2 and with h_{oe} still $100\,\mu S$, find the current gain.

    (a) 60   (c) 240
    (b) 120   (d) 75

4. The power gain with the circuit as in questions 2 and 3 is
(a) 20 dB   (c) 37·8 dB
(b) 20·8 dB   (d) 40·8 dB

5. A transistor has a value for α of 0·996. It is used as a common-emitter amplifier with a load of 2·5 kΩ. If h_{oe} is 100 μS estimate the current gain.
(a) 125   (c) 249
(b) 199   (d) 50

6. An emitter follower has an emitter load of 1 kΩ. If h_{fe} is 399 and h_{ie} is 1·5 kΩ find the input resistance.
(a) 1·5 kΩ (c) 401·5 kΩ
(b) 2·5 kΩ (d) 400 kΩ

7. Find the current gain in the circuit of question 6.

(a) 400   (c) 0·9975
(b) 150   (d) 200

8. Find the voltage gain in the circuit of question 6.
(a) 1   (c) 266
(b) 0·996   (d) slightly greater than 1

9. The power gain is
(a) 6 dB   (c) 22 dB
(b) 26 dB   (d) −0·03 dB

10. An FET has a g_m of 3 mA/V and an r_d of 20 kΩ.
It is used as a common source amplifier with a load of 5 kΩ. Draw the equivalent circuit and estimate the voltage gain.
(a) 3   (c) 15
(b) 60   (d) 12

11. The value of μ for the FET of question 10 (defined as for a valve) is
(a) 60   (c) 0·15
(b) 6·7   (d) μ has no meaning for a FET

12. Draw an equivalent circuit for the same amplifier using a constant-voltage generator.

13. From this equivalent circuit calculate the voltage gain.

5   Amplifiers II

5.1 Classes of bias

The amplifiers which have been considered in Chapter 4 were biased so
that they were what are called class A amplifiers. Class-A amplification
may be defined in two ways, both definitions, in fact, meaning the same.
First, it may be defined as the bias which places the quiescent point about
half-way along the usable part of the load line. This, of course, is just what
we set out to achieve in the design of the simple amplifiers which we have
discussed. Secondly, a class A amplifier may be defined as an amplifier in
which the collector or drain current flows at all times during the input-
signal cycle. This, again, is what we have been trying to achieve. It has
been pointed out that if the signal tries to drive the working point beyond
the bottom of the load line, the output will be clipped on the positive
peaks.

It might, at first, be thought that class-A operation is the only sensible
way in which to use an amplifying device. However, this is not so, and
mention will be made of other classes of bias, the ways in which they are
used being dealt with in later sections.

Class B

With class-B amplification the device is biased such that it only passes
current for half of the input cycle. In other words it is biased to cut-off.

Consider, as an example, the FET whose output characteristics together
with a load line, are shown in Fig. 5.1. With this load the FET is cut off
with a gate bias of -8 V. For class-A operation the quiescent point Q_1
would obviously be chosen. Biasing the FET at Q_2 would lead to class-B

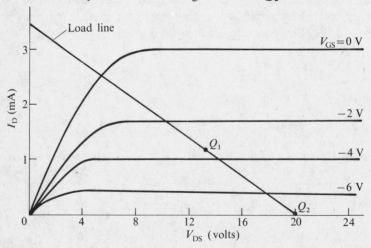

Fig. 5.1. Class-A and class-B bias.

operation. Only those parts of the input which are positive, i.e. which make the grid less negative than the bias of $-8\,V$, will cause drain current to flow, and the output will consist of negative half-cycles of voltage (negative because of the usual inversion produced).

The advantage of class-B operation is that with no input signal there is no current flowing and consequently no power being taken from the supply. This can be very important in a high-power circuit or where the supply is a dry battery. The other advantage of class-B operation is the increased efficiency—ideally 78·5 per cent.

Of course, such a saving in power would be pointless if the output were as grossly distorted as is implied. What is done is that two FETs or transistors are used, one amplifying the positive half-cycles and the other the negative. The results are then added together to produce a complete output waveform. This so called *push–pull* amplification will be dealt with in §5.5.

Class C

In the class C mode of operation the device is biased well beyond cut-off, so that current flows for less than half the input cycle, often for only a small fraction of a cycle.

The main use of this type of bias is in oscillator circuits and radio-frequency (r.f.) amplifiers where the output sine wave is produced by a parallel tuned circuit, the device passing just enough energy to overcome the losses in the resonant circuit.

5.2 Two-stage amplifiers

In the previous chapter we saw that a single common-emitter amplifier might give a power gain of 30 to 40 dB. Now, a system such as a record player needs a power gain of perhaps 8 000 000 times, or 69 dB, and clearly at least two stages of amplification will be required. This section will deal with some of the ways in which two stages may be coupled, or joined together.

R–C *coupling*

Consider the FET circuit of Fig. 5.2. The voltage output of T_1 is coupled to the gate of T_2 via a capacitor C_1, the purpose of this capacitor being to block the d.c. drain voltage of T_1 from the gate of T_2. The operation of the circuit is somewhat different from that of a single FET.

The a.c. load presented to T_1 is no longer simply R_{L1}. Remembering that the supply line and earth are the same as far as a.c. is concerned and that C_1 will be a short circuit to the frequencies of interest, R_{g2} is effectively in parallel with R_{L1}. The ability to see components as being in parallel to a.c., even though they are not in parallel as far as d.c. is concerned, is something which every electronic engineer must acquire.

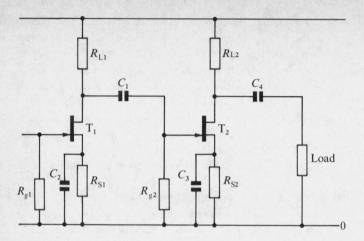

Fig. 5.2. Two-stage amplifier.

However, the d.c. load, which determines the d.c. load line and hence the operating point, is still R_{L1}, because C_1 is an open circuit to d.c. In other words the d.c. load is larger than the a.c. load.

Let us consider, as an example, an FET with the characteristics of Fig. 5.3, which is a repeat of Fig. 4.15. Let the load R_{L1} be 2 kΩ and the supply 20 V as in Chapter 4. The d.c. load line is drawn with the bias point Q (−2 V) marked. Now let us assume that R_{g2} is 250 kΩ. The a.c. load on the FET is 2 kΩ in parallel with 250 kΩ, which is a load of about 1·98 kΩ and the a.c. load line will be virtually the same as the d.c. one.

If the input signal is 2 V peak the drain current will swing from 0·5 mA to 8·4 mA or 7·9 mA peak to peak and the drain voltage from 19·1 V to 3·3 V or 15·8 V peak to peak, a voltage gain of 15·8/4 = 3·95 as was seen in Chapter 4.

The gain of the second stage of Fig. 5.2 is found in the same way, taking into account the load resistance connected to the output terminal. Let us assume that R_{L2} is, like R_{L1}, 2 kΩ and that the load is also 2 kΩ. The d.c. load line for T_2 is the same as that for T_1, but the a.c. load presented to T_2 consists of 2 kΩ in parallel with 2 kΩ, a load of 1 kΩ.

The current at the quiescent point is 4 mA and 4 mA in 1 kΩ produces a voltage of 4 V, so that the a.c. load line will cut the voltage axis at 16 V (12 V + 4 V). This line is also shown in Fig. 5.3. A 2-V peak input will now swing the drain current from 0·4 mA to 9·4 mA (9·0 mA peak to peak) and the drain voltage from 15·5 V to 7·5 V (8·0 V peak to peak). The voltage gain of T_2 is thus 8·0/4 = 2·0. The overall voltage gain is

$$3·95 \times 2·0 = 7·9.$$

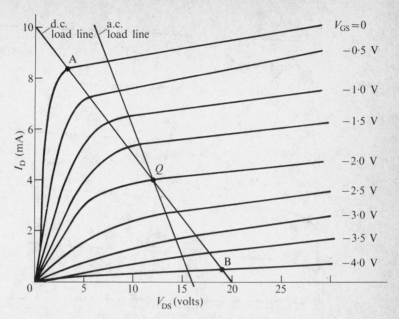

Fig. 5.3.

R–C coupling can also be used for transistors although in this case we are concerned with current gain as well as voltage gain. As an example we shall discuss the amplifier of Fig. 5.4.

The *d.c.* load applied to T_1 is $2 \, \text{k}\Omega$, consisting of the $1 \cdot 5$-$\text{k}\Omega$ collector load and the 500-Ω emitter resistor. For the circuit of Fig. 5.4 the load

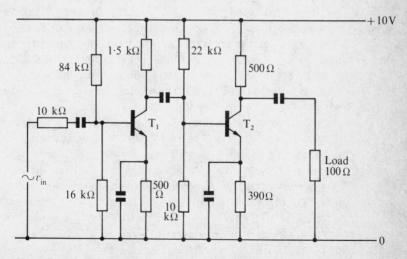

Fig. 5.4. Two-stage amplifier.

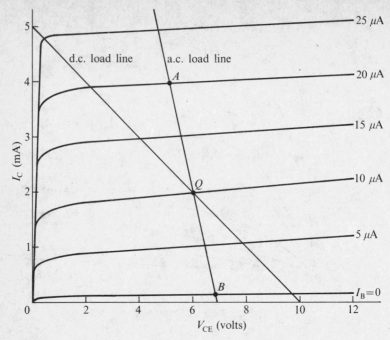

Fig. 5.5.

line is shown on the characteristics in Fig. 5.5. Now this circuit is what we have called a *constant collector-current* circuit. The base voltage is (16 × 10 V)/100 = 1·6 V and the emitter voltage is about 0·6 V less or 1·0 V. The emitter, and hence collector, current is 1 V/0·5 kΩ = 2 mA and the operating point Q has been marked on the diagram.

The *a.c.* load on T_1 is 1·5 kΩ in parallel with the input resistance of the second stage. (The 500 Ω emitter resistor is shorted out for a.c.)

h_{ie} for T_2 at the collector current it is taking can be found from the manufacturer's data. Let us assume that it is 700 Ω. The a.c. load on T_1 is thus 1·5 kΩ in parallel with 700 Ω both in parallel with the base bias resistors of 22 kΩ and 10 kΩ. The whole parallel combination amounts to 446 Ω.

An a.c. load line of this value can now be drawn through Q. Q is at 2 mA, 6 V. 2 mA in 446 Ω is 0·89 V, thus the line cuts the voltage axis at 6 + 0·89 = 6·89 V. It is almost vertical because the load is of a relatively low value. This means that the voltage gain is reduced. The current gain is very slightly increased, although not all of the output current will flow into T_2; some will flow into the d.c. load of T_1 and some into the biasing components of T_2. These three in parallel have a resistance of 1·23 kΩ, hence the fraction of the current flowing into the 700-Ω input resistance of T_2 is 1·23/1·93 = 0·64.

Fig. 5.5 shows that an input of 10 μA peak flowing into T_1 will produce an output current swing of (3·95 − 0·1) mA, giving a current gain of

$$A_i = \frac{3\cdot95 - 0\cdot1}{0\cdot02} = 192\cdot5.$$

The useful current gain, i.e. the current into T_2 divided by the input current will be $192\cdot5 \times 0\cdot64 = 123\cdot2$. Even this is not the true current gain because some of the generator current flows into T_1's two bias resistors, $84\,\text{k}\Omega$ and $16\,\text{k}\Omega$. These in parallel are $13\cdot4\,\text{k}\Omega$ and if h_{ie} of T_1 (at $2\,\text{mA}$) is say $2\,\text{k}\Omega$, the current into T_1 is $13\cdot4/15\cdot4$ or $0\cdot87$ of the generator current. Hence the effective current gain of T_1 is $0\cdot87 \times 126\cdot4 = 110$. The voltage swings from $5\cdot2\,\text{V}$ to $6\cdot8\,\text{V}$ or $1\cdot6\,\text{V}$ peak to peak.

Now the total resistance *seen* by the generator is $10\,\text{k}\Omega$ in series with ($84\,\text{k}\Omega$, $16\,\text{k}\Omega$, and $2\,\text{k}\Omega$) in parallel. This is $10 + 1\cdot7 = 11\cdot7\,\text{k}\Omega$. An input current of $20\,\mu\text{A}$ peak-to-peak requires a V_{IN} of $20\cdot10^{-6}\cdot11\cdot7\cdot10^3 = 0\cdot234\,\text{V}$ peak-to-peak, and the overall voltage gain of T_1 is $A_v = 1\cdot6/0\cdot234 = 6\cdot8$.

Fig. 5.6 shows the output characteristics of T_2. You should check that the quiescent collector-current of T_2 is about $6\cdot5\,\text{mA}$ and a d.c. load line of $890\,\Omega$ means a bias current of $50\,\mu\text{A}$.

The a.c. load on T_2 is $500\,\Omega$ in parallel with $100\,\Omega$ or $83\cdot3\,\Omega$ and if you draw an a.c. load line through your Q point you should find that for a base current input of $100\,\mu\text{A}$ peak to peak, the peak-to-peak swings of collector current and voltage are $12\cdot4\,\text{mA}$ and $1\cdot1\,\text{V}$ respectively. The current gain is thus $12\cdot4/0\cdot1 = 124$, the useful gain (into the $100\text{-}\Omega$ load) being $5/6$ of this or 103. If h_{ie} is $700\,\Omega$, the input voltage to produce a $100\text{-}\mu\text{A}$ base current is $700\,\Omega \times 100\,\mu\text{A} = 0\cdot07\,\text{V}$ and the voltage gain is $1\cdot1/0\cdot07 = 15\cdot7$.

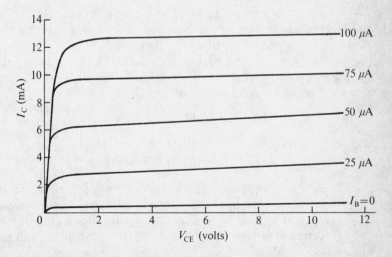

Fig. 5.6.

The overall current and voltage gains of the complete amplifier are

$$A_i = 110 \times 103 = 11\,330$$

$$A_v = 6.8 \times 15.7 = 107.$$

The overall power gain is

$$11\,330 \times 107 = 1\,212\,310 \quad \text{or} \quad 61\,\text{dB}.$$

This analysis has been carried out at some length because of the various important principles involved. In practice, an equivalent-circuit approach is much quicker, and this will now be demonstrated, assuming h_{fe} values for T_1 and T_2 of 200 and 130 respectively.

The equivalent-circuit current generator of T_1 will produce a current of $200\,i_b$ and, as explained above, 0.64 of this flows into T_2, i.e. there is an effective current gain of $200 \times 0.64 = 128$ for T_1. The total current gain of T_1 is thus $0.87 \times 128 = 111$ due to the effect of its bias resistors (see above).

The a.c. load on T_1 is $1.5\,\text{k}\Omega$, $22\,\text{k}\Omega$, $10\,\text{k}\Omega$, and $700\,\Omega$ in parallel, or $446\,\Omega$, so that the output voltage is $200\,i_b \times 0.446 = 89.2\,i_b$. i_b is 0.87 of the generator current i_{IN} so that the output voltage is $0.87 \times 89.2 \times i_{IN}$ $= 77.6\,i_{IN}$. The generator *sees* $11.7\,\text{k}\Omega$ so that $i_{in} = V_{in}/11.7$. Hence the output voltage of T_1, V_{o1}, is given by

$$V_{o1} = 77.7\,i_{in} = \frac{77.6\,V_{in}}{11.7},$$

and $A_v = \dfrac{77.6}{11.7} = 6.6.$

For T_2 the output current is $130\,i_b$ and $\frac{5}{6}$ of this flows into the load. Hence the current gain of T_2 is $\frac{5}{6}$ of 130, or 108 (the shunting effect of the bias resistors was taken into account in finding the current gain of T_1).

The output voltage of T_2, V_{o2}, is $130\,i_b \times 0.083 = 10.8\,i_b$. The input voltage is produced by i_b flowing into $700\,\Omega$, i.e. $0.7\,i_b$, and the voltage gain of T_2 is $10.8/0.7 = 15.4$.

These calculations give overall gains of

$$A_i = 111 \times 108 = 12\,000,$$

$$A_v = 6.6 \times 15.4 = 101.6,\text{ and}$$

$$A_p = 12\,000 \times 101.6 = 1\,219\,680 \quad \text{or} \quad 61\,\text{dB}.$$

These results compare well with those obtained from the characteristics.

These calculations are quite advanced, and likely to be considerably harder than is required of you at this level. However, they embody many fundamental principles, and it is considerably to your advantage to try to follow them.

Transformer coupling

Sometimes the load is coupled to the drain or collector via a transformer. This may be part of a tuned circuit designed to select a small band of frequencies (see §5.4) but here we will consider another application of transformer coupling. This is for matching, which will be needed if a device is feeding a load of resistance very different to its own output resistance. In this case a transformer can be used to alter the *apparent* resistance of the load.

Consider the transformer of Fig. 5.7. If it is 100 per cent efficient, the power on each side is the same:

$$V_1 I_1 = V_2 I_2.$$

But $V_2 = I_2 R_L,$

hence

$$V_1 I_1 = I_2^2 R_L$$

or $V_1 = \dfrac{I_2^2 R_L}{I_1}.$

The input resistance R_1 is, by definition, given by

$$R_1 = \frac{V_1}{I_1} = \frac{I_2^2 R_L}{I_1^2}$$

and as

$$\frac{I_2}{I_1} = \frac{N_1}{N_2},$$

(5.1) $$R_1 = \frac{N_1}{N_2}^2 R_L.$$

This means that the load R_L *appears* to be larger or smaller by the square of the turns ratio.

Early transistors had such a poor frequency response that they were nearly always used in the common-base configuration, which, as mentioned in Chapter 4, has a better frequency response than the other configurations. However, the low input resistance of one stage shorts out the very high output resistance of the previous stage and transformer coupling is used to *match* the two stages. This, of course, is costly, but illustrates one use of transformer coupling.

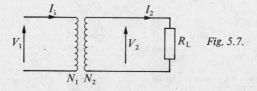

Fig. 5.7.

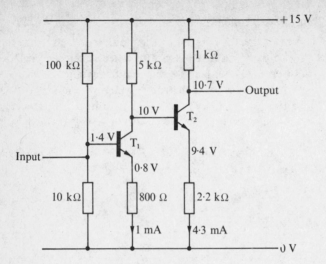

Fig. 5.8. Two-stage direct-coupled amplifier.

Another use is in matching the relatively high output resistance of a device into a low resistance load such as a loudspeaker. A full discussion of this will be given in Section 5.5.

Direct coupling

A consideration of the frequency response of amplifiers will be found in the next section, but it should be clear that neither of the previous coupling methods will operate in frequency down to d.c. Some applications require a response of this nature. The most obvious example is in the video amplifier of a television set, where a response from d.c. to a few

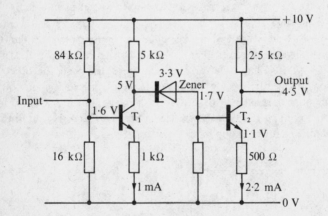

Fig. 5.9. Two-stage direct-coupled amplifier.

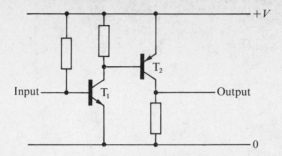

Fig. 5.10. *p–n–p/n–p–n
Direct-coupled amplifier.*

MHz is needed. There are also many examples in circuits designed for applications in mechanical engineering where frequencies are so low that the capacitances used in *R–C* coupling or the transformers in transformer coupling would need to be so large that these coupling methods become impracticable.

One obvious problem with *direct coupling* is that of d.c. levels. Consider the two-stage direct-coupled amplifier of Fig. 5.8. Here the emitter of T_2 is lifted to a relatively high voltage, requiring a higher voltage than would be required in an *R–C*-coupled circuit. Certain advantages are obtain by replacing the emitter resistor of T_2 by a zener diode. Another use of a zener diode in direct-coupled amplifiers is shown in Fig. 5.9, acting as coupling between T_1 and T_2. It drops a voltage but has negligible effect on *changing* voltages. Fig. 5.10 shows yet another circuit, this one employing both a p–n–p and an n–p–n transistor.

Many variations on these circuits are possible, a particularly useful one being the *Darlington pair*, two transistors connected together as shown in Fig. 5.11. This behaves as, and may be used as, a single transistor with an h_{fe} given by

(5.2) $$h_{fe} = h_{fe1} + h_{fe2}(1 + h_{fe1}),$$

where h_{fe1} and h_{fe2} are the current gains of T_1 and T_2 respectively. If T_1 and T_2 both had gains of 200 for example h_{fe} would be $200 + 200 \times 201 = 40\,4000$, equivalent to an α of 0·99998. Hence the circuit is sometimes called a *super-alpha pair*.

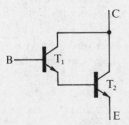

Fig. 5.11. *Darlington pair.*

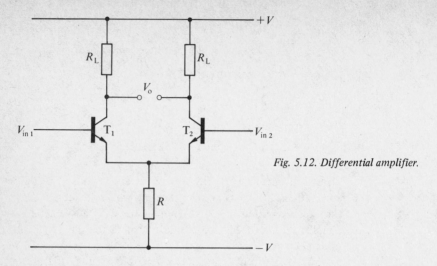

Fig. 5.12. Differential amplifier.

Differential amplifier

One problem suffered by most d.c. amplifiers is the problem of *drift*. A true d.c. amplifier will, of course, give zero output for zero input. The circuit of Fig. 5.8 could be altered to run off positive and negative supplies to enable this to be achieved. The problem then is that any disturbance of the d.c. conditions will alter the output. A slight change, for example, in the supply voltages will be amplified by both stages causing a significant output change. Similarly a change in temperature will alter the currents in the transistors and hence the output will drift away from zero for zero input. Clearly the effects on the first stage are the most important, being amplified by the rest of the circuit.

A circuit used to overcome these problems is the differential amplifier, Fig. 5.12. It is used as the first stage in nearly every high-gain d.c. amplifier, such as the integrated-circuit amplifiers we shall consider in Section 5.6.

In very simple terms the differential amplifier operates as follows. With both V_{in1} and V_{in2} zero the emitters, joined together, must be at about $-0.6\,\text{V}$. If $-V$ is relatively large, the current in R is about V/R ($0.6\,\text{V}$ being insignificant). By symmetry, this current divides equally between T_1 and T_2, so that the collectors will be at identical voltages and V_o will be zero.

If now V_{in1}, say, rises (V_{in2} still being zero), the current in T_1 will rise and that in T_2 must fall *by an equal amount* to keep the current in R constant at V/R. Thus the collector of T_1 falls and that of T_2 rises giving an output voltage. However, if something happens to affect T_1 and T_2 together, say a rise in $+V$ or $-V$, or a change in temperature, *both* collectors are affected in the same way and V_o will not alter. This circuit is thus not affected by changes affecting both transistors (common-mode signals) but does produce an output if there is a difference signal between the bases. (A difference amplifier is really a better name for the circuit.)

An example of the simple calculations involved is given in the Test questions 5.A. The differential amplifier is one of the important *building blocks* of electronics.

5.3 Frequency response

Even if an amplifier is designed to amplify as wide a range of frequencies as possible (a wide-band amplifier) it cannot amplify all frequencies from zero to infinity. If the coupling is via a capacitor it will not amplify down to zero frequency. The low-frequency response can be improved by making the capacitor as large as possible and, in the limit, direct coupling can be employed. However, there will always be a limit at the high-frequency end of the spectrum due to stray capacitance in parallel with the load, tending to short it out at high frequencies.

A wide-band amplifier will have a frequency response of the type shown in Fig. 5.13. Such responses are usually plotted with the gain in decibels against the log of the frequency (sometimes done by using logarithmic graph paper). A logarithmic frequency scale is used for many reasons, not least because the ear hears logarithmically and such a scale is therefore more meaningful.

By universal convention, the limits of the amplifier's response are the frequencies where the gain is 3 dB below the gain at *mid-frequencies*. The frequencies are usually denoted as f_1 and f_2 and the *bandwidth* of the amplifier is $f_2 - f_1$. Of course, the amplifier still possesses gain, perhaps considerable gain, outside this range, but this is, as stated above, the accepted definition of bandwidth. It is worth noting that a fall in gain of 3 dB is equivalent to a halving of power, and this is just discernable by the listener to an audio amplifier.

In most wide-band amplifiers f_2 is very high, often many MHz, and f_1 very low, a few Hz. In this case the bandwidth is effectively f_2. f_1 and f_2 are often called the lower and upper *3-dB frequencies* or the *half-power frequencies*.

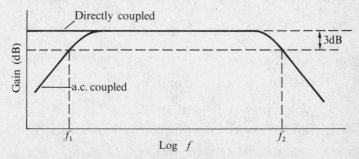

Fig. 5.13. Frequency response of an amplifier.

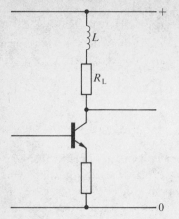

Fig. 5.14. High-frequency compensation.

One method which is commonly used to increase the value of f_2 is shown in Fig. 5.14. At frequencies where the gain is constant L has no effect (its value being chosen to this end). At frequencies where the stray capacitance starts to have an effect, the reactance of L begins to increase the load, to some extent offsetting the effect of C. The frequency response is improved a little, a typical response being shown in Fig. 5.15. Sometimes, as can be seen, the improvement is at the expense of a *bump* in the response.

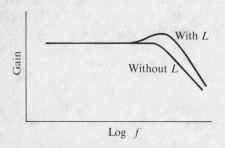

Fig. 5.15. Frequency response of circuit in Fig. 5.14.

5.A Test questions

1. Define the three types of bias, class A, B, and C
   (a) in terms of the cut-off of the device
   (b) in terms of the input (signal) cycle.

2. Give examples of applications of the three bias types of question 1.

3. Explain how class-B bias may be used to amplify a signal and discuss a major advantage of its use.

Fig. 5.16 shows a simple two-stage amplifier. T_1 *has an* h_{ie} *of 2 kΩ and* T_2 *an* h_{ie} *of 800 Ω. The* h_{fe}s *are 190 and 130 respectively and the* h_{oe}s *may be neglected. Both transistors are silicon, with* V_{BE}s *of 0·6 V. The output characteristics are given in Figs. 4.6 and 5.6 respectively. Questions 4–26 apply to this circuit.*

4. Draw the d.c. load line of T_1.

5. Calculate the value of R_1 required for a base bias current of 10 μA.

6. Calculate the value of R_2 such that the bias current of T_2 is 50 μA.

7. The input resistance of T_1 (i.e. as seen from C_1) is about (check that your value for R_1 is correct!)
    (a) 12 kΩ (c) 2 kΩ
    (b) 942 kΩ (d) 940 kΩ

8. The proportion of i_{IN} entering the base of T_1 is
    (a) 1   (c) 0·5
    (b) 0·002   (d) 0·2

9. The a.c. load presented to T_1 is
    (a) 604 Ω (c) 606 Ω
    (b) 2·5 kΩ (d) 2·47 kΩ

10. Draw an a.c. load line for T_1 through the quiescent point.

11. Estimate the current gain of T_1 (from base to collector).

12. The fraction of the output current of T_1 entering the base of T_2 is
    (a) 0·5   (c) all
    (b) 0·996   (d) 0·76

13. The overall current gain of T_1 (from v_{in} to the base of T_2 is
    (a) 141   (c) 70
    (b) 185   (d) 0·28

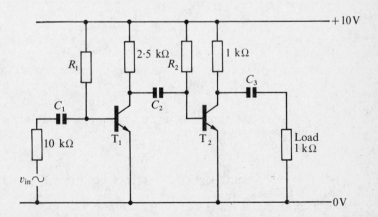

Fig. 5.16.

14. The a.c. load on T_2 is
    (a) 0·5 kΩ   (c) 1 kΩ
    (b) 2 kΩ   (d) zero

15. Draw the d.c. load line for T_2 onto Fig. 5.6.

16. Construct the a.c. load line through the quiescent point (50 μA bias).

17. Estimate the current gain of T_2 (from base to collector).

18. Estimate the current gain of T_2 into the load.

19. Calculate the overall current gain of the amplifier into the load.

20. Estimate the voltage gain of T_1 from v_{in} to the collector. (Hint—assume a reasonable value of i_{IN} say 20 μA peak to peak.)

21. Estimate the voltage gain of T_2, again assuming a reasonable value of input current, say 50 μA peak to peak.

22. Calculate the overall voltage gain.

23. Calculate the power gain in dB.

24. Using the concept of a current generator, estimate the current gain of the circuit.

25. Similarly estimate the overall voltage gain.

26. Estimate the power gain from the two previous results.

27. State two reasons for the use of a transformer for coupling two stages.

28. A transformer of 8 to 1 (step down) ratio has a load of 100 Ω across its secondary. If it is a perfect transformer, find the input resistance to the primary.

29. A load of 1000 Ω is connected across the secondary of a perfect transformer. If the primary has 50 turns and its input resistance is to be 10 Ω, the number of turns on the secondary is
    (a) 5   (c) 5000
    (b) 500   (d) 0·5

30. Explain two disadvantages of direct coupling.

31. Give two examples of the use of direct coupling.

32. A Darlington pair employs two transistors with h_{fe}s of 250 for T_1 and 200 for T_2. Find the effective value of alpha for the circuit.

33. In the differential amplifier circuit of Fig. 5.12 V is 15 V (and $-V$ is −15 V). If R is 10 kΩ and the base-emitter voltages are 0·6 V, the current in each transistor with both bases earthed will be
    (a) 1·5 mA   (c) zero
    (b) 1·44 mA   (d) 0·72 mA

34. If the quiescent voltage on each collector is to be 5 V with respect to earth, the value of R_L is
    (a) 14 kΩ   (c) 28 kΩ
    (b) 7 kΩ   (d) 2·5 kΩ

35. The value of V_O with both bases earthed is
    (a) 10 V   (c) 7·5 V
    (b) zero   (d) there is not sufficient information

36. If, with the base of T_2 still earthed, the input to T_1 is such that the collector current of T_1 rises by 0·25 mA find the change in collector current of T_2 assuming that the base-emitter voltages do not change.
    (a) 0·25 mA rise   (c) 0·5 mA fall
    (b) does not alter   (d) 0·25 mA fall

37. The voltage on the T_1 collector with respect to the collector of T_2 is now
    (a) 7 V   (c) −3·5 V
    (b) −7 V   (d) 14 V

38. Discuss the effects of temperature and supply voltage changes on the output of a differential amplifier.

39. An amplifier may be designed to operate down to zero frequency.
    (a) true
    (b) false

40. An amplifier may be designed to operate up to an infinite frequency.
    (a) true
    (b) false

41. An amplifier only has a gain greater than one if the output is not more than 3 dB below the gain at mid-frequencies.
    (a) true
    (b) false

42. Define bandwidth as applied to an amplifier.

43. The power gain of an amplifier is 60 dB. The frequencies f_1 and f_2 are the frequencies where the gain has fallen to
    (a) 30 dB   (c) 57 dB
    (b) 0 dB   (d) 42 dB

44. The mid-frequency power gain of an amplifier is 100 000. Calculate the power gain at f_1 and f_2.
    (a) 50 000   (c) 33 333
    (b) 99 997   (d) 70 700

45. An inductor is often added in series with a resistive load to increase the bandwidth of an amplifier.
Explain how the circuit operates and sketch a typical response.

5.4 Tuned amplifiers

A tuned amplifier is one designed to amplify a narrow band of frequencies. Tuned amplifiers are of fundamental importance in communication engineering. The early stages in a radio or television receiver are designed to select the required signal and reject others.

A simple amplifier designed to have a narrow bandwidth uses a tuned circuit as its load (Fig. 5.17). A graph of the impedance of the circuit against frequency is shown in Fig. 5.18. Z_m is the impedance at resonance which is given by

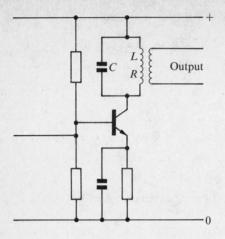

Fig. 5.17. RF amplifier.

(5.3)
$$Z_m = \frac{L}{CR}.$$

The approximate value of f_0 is given by

(5.4)
$$f_0 = \frac{1}{2\pi\sqrt{(LC)}}.$$

The bandwidth is $(f_2 - f_1)$, f_1 and f_2 being the frequencies where Z_m is reduced to $Z_m/\sqrt{2}$, which is when the output voltage will have dropped by 3 dB. As the gain is proportional to the collector load, this curve is also a graph of gain against frequency.

The quality of a tuned circuit is sometimes quoted in terms of its *Q factor* or *selectivity*. Q_0 is given by

(5.5)
$$Q_0 = \frac{\omega_0 L}{R} = \frac{1}{\omega_0 C} = \frac{1}{R}\sqrt{\frac{L}{C}},$$

where $\omega_0 = 2\pi f_0$.

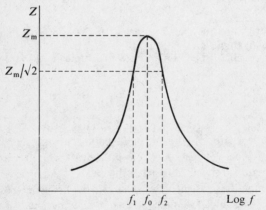

Fig. 5.18. Frequency response of tuned amplifier.

Bandwidth and selectivity (the ability to *select* a frequency) are related and depend on f_0. (Consider the relative effectiveness of a 1-kHz bandwidth if f_0 is 2 kHz or if it is 2 MHz!) The relationship is

(5.6)
$$Q_0 = \frac{f_0}{\text{bandwidth}}.$$

Thus a tuned circuit with an f_0 of 1 MHz and a bandwidth of 10 kHz has a Q_0 of 100, as does a circuit with an f_0 of 1 kHz and a bandwidth of 10 Hz.

As an example consider a coil of inductance $100\,\mu H$ and resistance $10\,\Omega$ in parallel with a capacitor of 100 pF.

$$Z_m = \frac{10^{-4}}{10 \times 10^{-10}} = 10^5\,\Omega,$$

$$f_0 = \frac{1}{2\pi\sqrt{(10^{-4} \times 10^{-10})}} = 1\cdot59\,\text{MHz},$$

$$Q_0 = \frac{1}{R}\sqrt{\frac{L}{C}} = \frac{1}{10}\sqrt{\frac{10^{-4}}{10^{-10}}} = 100,$$

therefore the bandwidth

$$= \frac{1\cdot59}{100}\,\text{MHz} = 15\cdot9\,\text{kHz}.$$

If the resistance of the coil were, say, doubled, f_0 would not alter appreciably, Q_0 would become 50 and the bandwidth double, becoming $31\cdot8$ kHz. Z_m, and hence the gain, which is proportional to Z_m, would halve.

The output of the circuit of Fig. 5.17 could be taken from the collector, but is more often taken from a secondary winding coupled to the coil as shown.

Frequently a relatively wide bandwidth is required, for example 5 or 6 MHz for a colour television signal. To achieve bandwidths of this order both the primary and secondary of the transformer are tuned, Fig. 5.19, each to a slightly different frequency, leading to a response such as the one shown in Fig. 5.20.

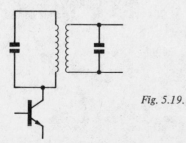

Fig. 5.19.

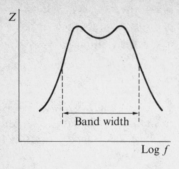

Fig. 5.20. Frequency response of circuit in Fig. 5.19.

5.B Test questions

1. Discuss the uses of tuned amplifiers.

2. A tuned circuit, used in a tuned amplifier, has a coil of resistance $5\,\Omega$ and inductance $100\,\mu H$. The value of capacitance needed to resonate at 1 MHz is

    (a) 1·59 pF   (c) 2·53 μF

    (b) 15·9 pF   (d) 253 pF

3. The impedance of this tuned circuit, at resonance, is

    (a) 79 kΩ (c) zero

    (b) 79 Ω (d) $0{\cdot}5 \times 10^{-6}\,\Omega$

4. The value of Q_0 is

    (a) 126   (c) 79

    (b) 100   (d) very small

5. The bandwidth of the circuit is

    (a) 1 MHz   (c) 0·000126 Hz

    (b) 7·9 kHz   (d) zero

6. The voltage gain of the circuit at 1 MHz is 50. The voltage gain at 1·0079 MHz will be

    (a) 35·4   (c) 25

    (b) 47   (d) 70·7

7. At what other frequency will the gain be the same as that in question 6?

8. The amplifier using the tuned circuit of the previous questions needs a bandwidth four times as large. How could the resistance of the coil be altered to produce this bandwidth?

9. What would happen to the gain in this case?

10. Discuss how the bandwidth of an r.f. amplifier could be increased without this change in gain.

5.5 Large-signal amplifiers

It has been emphasized many times that all amplifiers amplify power. Nevertheless, the circuits with which we have been dealing so far operate at very low power levels.

Many amplifiers need, ultimately, to produce a relatively large power output, and the final stage of such a circuit is designed to this end, being sometimes called a *power amplifier*. As the device being used will normally be operated over a large part of its characteristics, a better name is a *large-signal amplifier*.

The application which comes most readily to mind is the stage feeding the loud-speaker in a radio set or in hi-fi equipment. Loudspeakers have, for various reasons, relatively low resistances, 4, 8, and 16 ohms being standard values. Many transistors have output resistances much higher than this and this presents matching problems.

Transformer coupling may be used as mentioned in §5.2. A transistor coupled to a load via a transformer is shown in Fig. 5.21. Fig. 5.22 shows the characteristics of a power transistor, i.e. one designed to give a large output power. Note that much higher base currents than we have been used to are involved.

The d.c. load line will be vertical, as shown, because the transformer's primary resistance may be considered negligible. A suitable bias would be 15 mA with, as can be seen, a quiescent collector current of 1·14 A.

A suitable a.c. load line is shown and represents a load of 20 Ω. As the actual load is 1 Ω this requires a transformer turns ratio of $\sqrt{20}$, or 4·47 to 1 (N_p to N_s). The power dissipated in the transistor will be 10 V × 1·14 A, which is the d.c. power supplied. If a peak signal of 5 mA is now applied the output swings from A to B, a swing of 16·0 − 5·3 = 10·7 V

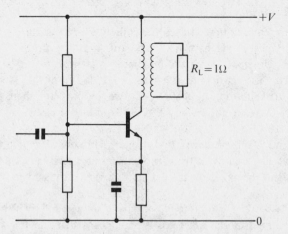

Fig. 5.21. Large signal amplifier.

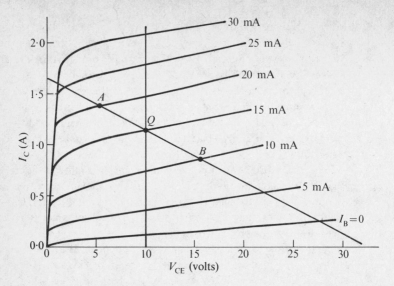

Fig. 5.22. Power transistor characteristics.

and $(1 \cdot 37 - 0 \cdot 84) = 0 \cdot 53$ A. The output power is found by dividing the product of these peak-to-peak values by 8 (i.e. $2\sqrt{2}$ twice) giving $(0 \cdot 53 \times 10 \cdot 7)/8 = 0 \cdot 71$ W.

It can be seen that the voltage waveform is not symmetrical, $6 \cdot 0$ V positive peak and $4 \cdot 7$ V negative peak. This is, in fact, distortion containing *even* harmonics (second, fourth, etc.), which is one of the problems of this particular circuit. Another difficulty is the fact that it is not desirable to have d.c. flowing in the transformer primary. Further, the power being dissipated even with no signal is quite large ($11 \cdot 4$ W).

Push–pull circuit

A circuit overcoming these difficulties is the *push–pull* circuit. The perfect push–pull amplifier will have no *even* harmonic distortion, the only distortion being the small amount of odd harmonics present.

The circuit uses two transistors (or valves) fed from signals which are identical but $180°$ out of phase. The two outputs are then added to produce a single output signal. A typical push–pull circuit is shown in Fig. 5.23.

Transformer Tr1 produces two signals $180°$ out of phase, and the amplified versions of these flow in opposite directions in the primary of Tr2, the signal being reconstituted in the secondary and hence in the load.

Whilst class-A bias can be used, certain advantages follow if the transistors are operated in class B. Class-B bias means that the transistors are biased to cut-off so that each only amplifies one half-cycle of the input, the halves being added to form a whole in the transformer secondary. The main advantage is that, unlike class A, the power taken from the supply decreases with decreasing input signal, becoming zero with no input. This leads to far greater efficiency, a very important factor in modern battery radio sets.

One disadvantage of class B is that, with silicon transistors, about 0·6 V is needed before they conduct, leading to a *dead space*, producing what is known as *cross-over distortion*. In practice a bias not quite sufficient to cut-off the transistors is used, eliminating this form of distortion.

The circuit of Fig. 5.23 can be improved by the omission of Tr2 thus reducing cost (and distortion), Fig. 5.24.

The half-cycles of signal are shown marked 1 and 2. T_1 conducts on half-cycle 1, passing current in the load in the direction shown. During half-cycle 2, T_2 conducts. The current in the load, as can be seen, is alternating.

This circuit still requires one transformer, or some other means, of producing the two out-of-phase signals. However this can be avoided if one of the n-p-n transistors is replaced with a p-n-p type as shown in Fig. 5.25, the circuit being known as a complementary pair output stage. To obtain the advantage of reduced distortion it is essential that the two transistors be identical, except that one is p-n-p and the other n-p-n. The circuit has the disadvantage of requiring two power supplies (+ and −) and a circuit using only one is given in Fig. 5.26. The operation is similar to the previous circuit except that C, a very large capacitor, charges

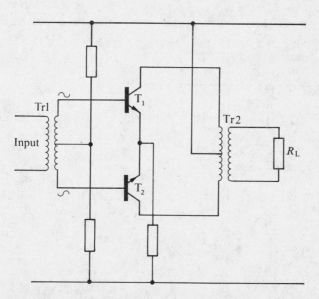

Fig. 5.23. Push-pull amplifier.

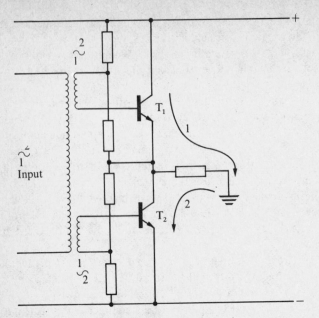

Fig. 5.24. Transformerless output stage.

with the polarity shown during the positive half-cycle and on the negative half-cycle it discharges, sending current through T_2 and the load.

Parasitic oscillation

Some amplifiers may, if incorrectly designed or constructed, oscillate. Oscillations are due to feedback, as will be seen in a later chapter. Great care must be taken in the layout of amplifiers to avoid such parasitic oscillations.

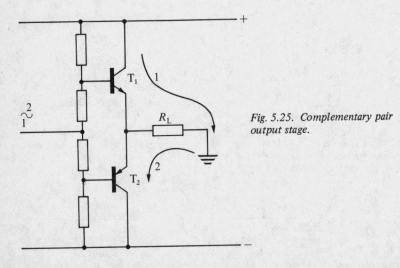

Fig. 5.25. Complementary pair output stage.

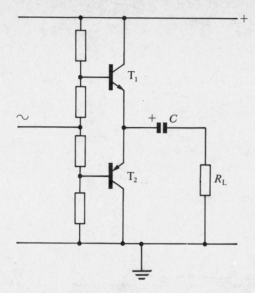

Fig. 5.26. Complementary pair output stage with one power supply.

5.C Test questions

1. Discuss the differences between small- and large-signal amplifiers.

A silicon transistor with characteristics given in Fig. 5.27 operates in the circuit of Fig. 5.28. Questions 2–5 are based on this circuit.

2. If the base bias current is 15mA, find the quiescent power taken from the supply.

3. Draw an a.c. load line.

4. Estimate the largest reasonable voltage and current swings.

5. Calculate the output power for the conditions of question 4.

6. Estimate a suitable value for R_1 to provide a bias of 15 mA.

7. What are the advantages of push–pull operation?

8. Explain why transistors are not often operated exactly in class B.

9. Sketch the circuit of a push–pull amplifier employing two p-n-p transistors and with no output transformer.

10. Draw the circuit diagram of a complementary pair output stage
    (a) with two power supplies
    (b) with one power supply
Explain the operation of each.

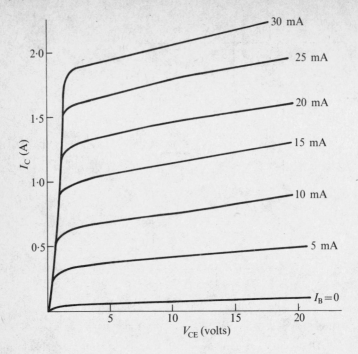

Fig. 5.27.

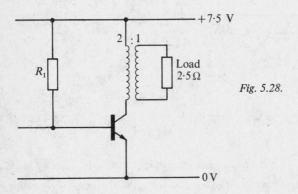

Fig. 5.28.

5.6 Linear integrated circuits

There is not the space in this book to deal in any way with the manufacture and construction of integrated circuits (ICs).

An IC is a complete electronic circuit (including wiring) manufactured on a minute chip of silicon. Although the chip will be perhaps $1\frac{1}{4}$ mm square and $\frac{1}{4}$ mm thick, it is manufactured in a package somewhat larger. Various standard packagings are in use, a typical one being about 19 × 7 × 3 mm with 14 pins.

Over the last decade or so the price of ICs has fallen dramatically. At the same time their complexity in terms of the number of circuit elements per chip has risen considerably.

There are two main catagories of IC—linear and digital. The latter will be mentioned in Chapter 8.

The basic linear IC uses circuit configurations of the type which have been discussed—d.c. coupled stages, Darlington pairs, differential amplifiers, etc.

Numerous circuits are available in integrated form. These include wideband amplifiers, audio amplifiers, r.f. amplifiers, voltage regulators, operational amplifiers. The operational amplifier, as will be seen in the next chapter, is a high-gain d.c. amplifier. A typical, cheap operational amplifier is the type 702. It consists of about 9 transistors and 11 resistors (the actual circuit varies slightly from one manufacturer to another) and it has a voltage gain of about 4000. This is with supplies of $+12\,V$ and $-12\,V$ when it dissipates a total power of less than $100\,mW$.

At the time of writing (1978) some typical prices for linear ICs are

Simple operational amplifier	36p
FET input operational amplifier (input resistance $10^{14}\,\Omega$)	£2.50
Complete audio amplifier	£2.00

The advantages of ICs speak for themselves—low cost, small size, low power consumption, ease of servicing, and reduction in maintenance costs.

6   Feedback

6.1 Introduction

Feedback is a subject on its own. It concerns the feeding back of information from the output to the input of a system to modify the input so that the output is controlled in a particular way.

If this sounds rather complicated, consider a simple electronic example. It is often desirable to maintain the gain of an amplifier constant, but the latter may change because of changes in transistor characteristics, changes of temperature etc. It would be possible for a human operator to watch the output and if the gain were to fall to alter a gain control to counteract it. This is a type of feedback system, but we usually think of feedback more in terms of an automatic system. In such a system the output controls the input so that the gain stays constant. If the gain falls, more signal is applied to the input so that the output tries to maintain its initial value.

Although we shall only be considering electronic applications, feedback theory is applied in many other areas. Feedback may be used to control the speed of a motor or to maintain the thickness of steel being made in a rolling mill. The state of a nation's economy (the output) will change the supply of money (the input) which will affect the economic situation.

The feedback equation

Feedback may be applied to an amplifier to control voltage or current gain. As an example we will consider a voltage amplifier with a gain of A_v. Figure 6.1 shows a diagrammatic representation of such a system.

Obviously

(6.1)
$$v_o = A_v v_{in}$$

and with v_{in} constant, any change in A_v will directly change v_o.

Note that although earth has been shown in Fig. 6.1 it is usually omitted from such diagrams, all voltages being assumed to be with respect to earth.

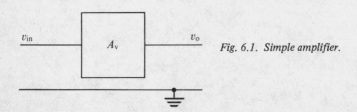

Fig. 6.1. Simple amplifier.

In a feedback system a fraction β of the output is *fed back* and *added* to the input signal. This is shown in Fig. 6.2, the sum of the input and fedback voltages being v_i. The fraction is often produced by a simple potential divider across the output as will be seen later in Fig. 6.5. Now,

(6.2) $$v_i = v_{in} + \beta v_o.$$

The gain of the amplifier itself is still, of course, A_v and hence

(6.3) $$v_o = A_v v_i,$$

(6.4) or $$v_i = \frac{v_o}{A_v}$$

Substituting eqn (6.4) in eqn (6.2)

$$\frac{v_o}{A_v} = v_{in} + \beta v_o$$

$$v_o = A_v v_{in} + A_v \beta v_o$$

$$v_o - A_v \beta v_o = A_v v_{in}$$

(6.5) $$v_o (1 - A_v \beta) = A_v v_{in}$$

The gain of the whole system is v_o/v_{in} and this is often given the symbol A_{vf} meaning the voltage gain *with* feedback.

(6.6) $$A_{vf} = \frac{v_o}{v_{in}} = \frac{A_v}{1 - A_v \beta} \quad \text{(from eqn (6.5))}.$$

This is one of the most important equations in electronics.

Now $A_v \beta$ might be positive or negative and four cases will be considered.

(a) $A_v \beta$ *positive and less than one.* In this case $(1 - A_v \beta)$ is positive and less than one and hence A_{vf} is greater than A_v. The feedback has increased the gain. We have, in effect, added part of the output and reamplified it. This is called *positive feedback* and although it seems a good idea, it has many disadvantages which will be mentioned later.

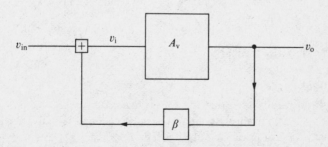

Fig. 6.2. Feedback loop.

(b) $A_v\beta$ *negative.* This occurs if either A_v or β (but not both) are negative. A negative gain means $180°$ phase shift. Many amplifier stages, as we have seen, produce a $180°$ phase shift and most amplifiers with an *odd* number of stages have such a phase shift. If the amplifier has zero phase shift then β can be made negative by *subtracting* $A_v\beta$ from the input.

It should be noted here that some books (and lecturers) assume, at the start, that βv_0 is subtracted from the input and this results in eqn (6.6) having $(1 + A_v\beta)$ as denominator. This method of approach will give the same answer, as will be demonstrated in example 1 (a) following §6.2. However it is vital that you do not confuse the two methods.

If, then, $A_v\beta$ is negative, the term becomes positive in eqn (6.6), to make A_{vf} *smaller* than A_v. It might at first sight seem a bad idea to *reduce* the gain, but many advantages follow the application of *negative feedback*, as it is called, and these will be discussed in §6.2.

(c) $A_v\beta = 1$. This could occur, for example, if A_v were 100 and β 1/100. Eqn (6.6) then gives

$$A_{vf} = \frac{A_v}{0}.$$

$A_v/0$ is infinity, which means that we get an output even with no input! If you do not like to think of gains of infinity, you may prefer to think in terms of the feedback voltage being just sufficient to maintain the output with no input signal.

It is possible to design circuits such that $A_v\beta$ is unity at one and only one frequency. The output is then sinusoidal and the circuit is a sine-wave oscillator. Oscillators will be the subject ot the next chapter.

(d) $A_v\beta$ *positive and greater than one.* In this case the circuit will, like (c) be unstable.

Lastly, before we consider the effects of negative feedback, three definitions will be given. $A_v\beta$ is called the *loop gain.* It is the gain from the input through the amplifier and feedback network. $(1 - A_v\beta)$ is called the feedback factor. The amount of feedback is often quoted in decibels as

(6.7) $$\text{dB of feedback} = 20\log_{10}\left|\frac{A_{vf}}{A_v}\right| = 20\log_{10}\left|\frac{1}{1 - A_v\beta}\right|.$$

6.2 The effects of negative feedback

Gain

We have already seen that the gain is *reduced* by negative feedback. As an example, if the gain without feedback of an amplifier is 1000 with a $180°$ phase shift and $\frac{1}{10}$ of the output is fed-back then

$$A_v = -1000 \text{ (the minus meaning } 180° \text{ shift)}$$

$$\beta = 0 \cdot 1$$

$$A_v\beta = -1000 \times 0 \cdot 1 = -100$$

$$1 - A_v\beta = 1 - (-100) = 1 + 100 = 101$$

$$A_{vf} = \frac{-1000}{101} = -9 \cdot 901.$$

This fall in gain is the big disadvantage of negative feedback, the price we have to pay for the numerous advantages discussed below. It means that to produce a certain gain, a negative feedback amplifier must employ more transistors or valves than one without feedback.

Stability of gain

The first, and most obvious, advantage of negative feedback is that of stability of gain. The gain of an amplifier might slowly change with time, due to ageing of components or other factors. It might change suddenly if the transistor is replaced owing to failure. Consider, for example, that the gain of the amplifier dealt with above falls to -500. This 50 per cent fall might not be too important, say, in a radio set where the listener would compensate for it by adjusting the volume control. However in an amplifier used as part of a measuring system it would be catastrophic!

Now β is, we can assume, still one-tenth, because it is probably determined by a pair of resistors forming a potential divider. Hence

$$A_v = -500$$

$$A_v\beta = -50$$

$$1 - A_v\beta = 1 - (-50) = 51$$

$$A_{vf} = \frac{-500}{51} = -9 \cdot 804.$$

This is only about a 1 per cent fall in gain.

Whilst these arithmetical calculations are very easy to follow, you might prefer a more physical explanation of how the feedback operates in keeping the gain almost constant.

With $A_v = -1000$ the voltages at various points in the circuit, assuming an input voltage of unity, are shown in Fig. 6.3. At the input $-0 \cdot 9901$ is added to 1 giving $0 \cdot 0099$, which, when amplified -1000 times, gives $-9 \cdot 9$ at the output. (The actual output is $-9 \cdot 901$ the difference being because we have not worked to enough places in the calculations.)

Fig. 6.4 shows the situation when A_v has fallen 50 per cent. Because the output has fallen, the voltage fed-back is less and the input to the actual amplifier, v_i, is larger. It can be seen that the circuit is very cunning, increasing the input to the amplifier if the gain falls!

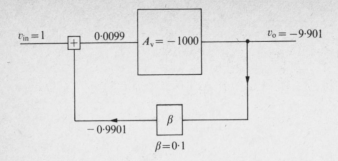

Fig. 6.3.

Another way of looking at this stability of gain is to consider the case when the magnitude of $A_v\beta$ is much larger than one. The magnitude of $A_v/(1 - A_v\beta)$ is then approximately equal to

$$\frac{A_v}{A_v\beta} = \frac{1}{\beta}.$$

Both of the calculations above have a β of 0·1 and hence the gain should be about 1/0·1 or 10. Accurate calculations gave 9·901 and 9·804. If A_v had been, say, 10^6 and β 0·1.

$$A_{vf} = \frac{-1\,000\,000}{1 + 100\,000} = -9\cdot9999$$

which is, indeed, very close to —10.

We can see that an amplifier with a predictable and very stable gain can be designed using negative feedback. We need not worry about the large variations in h_{fe} of transistors for example. In fact the gain of a negative feedback amplifier can be made to depend almost entirely on the value of β (as long as $A_v\beta$ is large) and β itself may only depend on two resistors, as shown in Fig. 6.5.

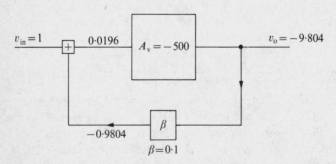

Fig. 6.4.

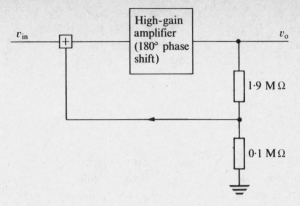

Fig. 6.5.

In this circuit

$$\beta = \frac{0 \cdot 1}{1 \cdot 9 + 0 \cdot 1} = 0 \cdot 05.$$

Hence the gain will be very close to $1/0 \cdot 05 = 20$ and this has been determined with no reference to the amplifier itself except to know that the gain is very high and that there is a $180°$ phase shift.

Bandwidth

We have just seen that the gain of a negative-feedback amplifier stays almost constant if the value of A_v, the gain without feedback, falls. This fall may, of course, be due to any cause and in particular to the drop in gain we expect at high and low frequencies due to stray and coupling capacitances respectively.

The gain with feedback tends to stay constant as long as the product $A_v \beta$ is large compared to one. Typical responses are shown in Fig. 6.6,

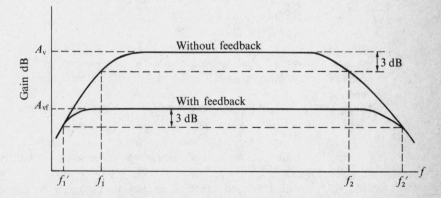

Fig. 6.6. Improved bandwidth.

f_1 and f_2 being, as usual, the 3-dB frequencies of the amplifier without feedback and f_1' and f_2' the 3-dB frequencies with feedback.

It can be shown that f_1 is reduced by a factor $(1 - A_v\beta)$ and f_2 is increased by the same factor. Remember that with negative feedback $(1 - A_v\beta)$, the feedback factor, is greater than one because $A_v\beta$ is negative.

(6.8)
$$f_1' = \frac{f_1}{1 - A_v\beta},$$

(6.9)
$$f_2' = f_2(1 - A_v\beta).$$

As the bandwidth of a wide-band amplifier is almost equal to f_2 (since f_2 will be very much larger than f_1) the bandwidth is increased by the feedback factor. As the gain has been reduced by the same factor, the product of gain and bandwidth for a wide-band amplifier is, in fact, constant. This so called gain–bandwidth product is a very important quantity for a wide-band amplifier. It enables us to find the gain we can obtain for a given bandwidth or vice versa. Gain is a ratio so the gain–bandwidth product is measured in Hz.

Distortion

Most amplifiers consist of a number of small-signal stages followed by an output or large-signal stage. Almost all the distortion produced in a well-designed amplifier will be in this output stage, where we are tending to depart from the linear part of the characteristics.

Obviously if negative feedback were applied to the amplifier without changing the input signal, the signal levels would fall so much that distortion would be almost non-existant. If, however, the input signal is increased to keep the output the same with feedback as it was without, it can be shown that distortion is reduced by $(1 - A_v\beta)$. Thus we can obtain a very considerable reduction in distortion.

The comments of the previous paragraph are very often confused, some sources actually saying that because distortion is only reduced by the same amount as gain there is no net improvement! It must be emphasized that there is a great reduction in distortion as long as the output is maintained constant by increasing the input signal. This increase in input signal is carried out *before* the feedback loop. The preamplifier producing the increase is operating at such a low signal level that the distortion it adds may be neglected. Fig. 6.7 shows such an arrangement.

In hi-fi circuits the preamplifier, as well as providing gain, is normally used to carry out the frequency compensation necessary for playing records and tapes and also incorporates the tone and volume controls.

One thing must be realized concerning the reduction of distortion by negative feedback. It only reduces distortion occurring in the main amplifier or feedback loop and will have no effect whatsoever on distortion entering as part of the signal. Some signals, such as those from 'pop' records, have 'distortion' built in on purpose and the listener would hardly expect this 'distortion' to be reduced by the amplifier.

It is interesting to consider just how the feedback does reduce distortion. If the input signal is, for example, a sine wave but leaves the amplifier, as v_o, distorted, this distorted wave is fed back out of phase and predistorts the input signal in such a way that the distortion produced by the amplifier tries to turn it back into a sine wave.

Noise

This is a rather more difficult problem. Some authorities state that it, like distortion, is reduced by $(1 - A_v \beta)$. This is, in fact, not quite true because of the random nature of noise. Further, the preamplifier required to increase the gain will introduce a considerable amount of noise (unlike distortion) and it is even possible to end up with more noise! In general, however, a well-designed negative-feedback system will reduce noise (produced in the amplifier or feedback loop, of course).

Noise which is not random such as 50- or 100-Hz mains hum will be reduced by $(1 - A_v \beta)$ but again more might be introduced in the pre-amplifier.

Input and output resistances

The importance of the input and output resistances of an amplifier has been emphasized before. As explained, these parameters are as important as gain itself. Feedback allows both the input and output resistances to be increased or decreased, depending on the circuit used.

First consider the input circuit. The actual addition of v_{in} and βv_o may be done in two ways. Either they can be added in series or in parallel, giving series or shunt feedback. Fig. 6.8 shows the two methods in block diagram form.

As long as the feedback is negative, series feedback increases the input resistance. The feedback voltage opposes v_{in}, because it is negative feedback, so that v_{in} supplies less input current with feedback and thus sees a higher input resistance. Shunt feedback reduces the input resistance because current from v_{in} flows not only into the amplifier but also back into the feedback network, so that v_{in} supplies more current with feedback and so sees a lower input resistance.

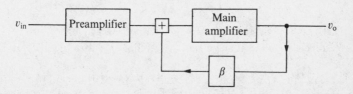

Fig. 6.7. Amplifier and preamplifier.

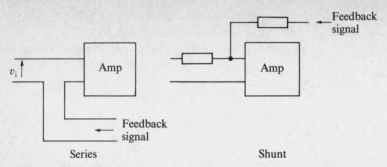

Fig. 6.8. Types of feedback.

The output of the amplifier can supply a feedback signal in one of two ways. It can produce feedback proportional to the output voltage (voltage feedback) or proportional to the output current (current feedback). Voltage feedback reduces the output resistance because it tries to keep the output voltage constant, that is to make the output look like a voltage generator and a voltage generator should have a low output resistance (preferably zero). Similarly current feedback tries to keep the output current constant, and as a current generator should have a high resistance its effect is to increase the output resistance.

There are, then, four possibilities, and a summary of the effects of negative feedback is given in Table 6.1.

Table 6.1. Effects of negative feedback

Type	Gain	Gain stability	Distortion	Noise	R_{in}	R_{out}
Voltage/series	Reduced	Improved	Reduced	Reduced	Increased	Decreased
Voltage/shunt	Reduced	Improved	Reduced	Reduced	Decreased	Decreased
Current/series	Reduced	Improved	Reduced	Reduced	Increased	Increased
Current/shunt	Reduced	Improved	Reduced	Reduced	Decreased	Increased

We can thus control the input and output resistances in either direction by choosing the feedback circuit which we use. It is even possible to obtain the many advantages of feedback without altering these resistances by a suitable mixture of types of feedback.

Stability

Lastly attention must be drawn to one of the dangers of applying negative feedback to a multistage amplifier, that is the danger of it becoming unstable or oscillating.

Any amplifier stage has a finite bandwidth. Even if the frequency response extends down to zero frequency there will be an upper limit determined by the stray capacitances in the circuit. This will not only introduce a fall in gain but also a change in the phase shift produced, so that it departs from 180°. Whilst a normal common-emitter, common-source, or common-cathode stage could not produce a phase shift of zero, a three-stage amplifier, normally giving 180° shift, can. This means that what was negative feedback can, at some frequency, become positive. If, at this frequency, the loop gain is one, then $A_v \beta = 1$ and

$$A_{vf} = \frac{A_v}{1 - A_v \beta} = \frac{A_v}{0} \text{, or infinity,}$$

and the circuit will oscillate as explained earlier.

Any feedback system may become unstable. The steel rolling mill or the nation's economy could produce dangerous results if instability were allowed to occur.

Clearly it is essential in designing a negative-feedback system to avoid this situation. A study of how to predict whether a feedback system will oscillate, and what measures to take to prevent it is rather complicated and well beyond the scope of this book.

Chapter 7 deals with circuits which are deliberately designed to be unstable, that is oscillators, and such circuits are, of course, designed to have a value of $A_v \beta$ equal to one at the frequency of oscillation.

Positive feedback

The effects of negative feedback considered above are reversed with positive feedback. Thus, although the gain is increased, it is less stable (i.e. it depends more on the gain of the actual amplifier), distortion and noise are increased and bandwidth is reduced. Also, of course, there is far more danger of oscillation.

Whilst positive feedback is used in certain applications (as well as oscillators) these dangers must always be borne in mind.

Examples

Example 1

An amplifier has the following properties:
voltage gain = 500 with 180° phase shift.
f_1 = 50 Hz
f_2 = 30 kHz
total harmonic distortion = 5 per cent.
Voltage series feedback is applied with a β of 0·2
(a) Find the voltage gain with feedback.
(b) Determine the percentage fall in gain with feedback if the amplifier gain falls by 20 per cent.
(c) Find f_1' and f_2'.
(d) What is the gain-bandwidth product with and without feedback?

(e) Find the total harmonic distortion (with feedback) assuming that the output remains constant.

(f) What is the loop gain?

(g) What is the feedback factor?

(h) Find the amount of feedback applied in dB.

(i) Find the required gain of the preamplifier to maintain the output constant when the feedback is applied.

Solution

(a) $A_v = -500$ $A_v\beta = -500 \times 0.2 = -100$ $1 - A_v = 101$

$$A_{vf} = \frac{-500}{101} = -4.95$$

Note

As explained earlier some books use

$$A_{vf} = \frac{A_v}{1 + A_v\beta}$$

assuming that the fed-back voltage is subtracted.

In this case A_v would be considered as $+500$ (the minus having been assumed in deriving the formula) and

$$A_{vf} = \frac{500}{1 + 500 \times 0.2} = 4.95 \text{ as before.}$$

The author considers that whilst both approaches have merit, taking all factors into account, eqn (6.6) is the better method to use. It is essential to emphasize that the two methods must not be confused.

(b) $500 - (20\% \text{ of } 500) = 400$ $A_v = -400$

$$A_v\beta = -400 \times 0.2 = -80$$

$$1 - A_v\beta = 81$$

$$A_{vf} = \frac{-400}{81} = -4.94$$

$$\text{percentage fall} = \frac{4.95 - 4.94}{4.95} \times 100 = 0.2 \text{ per cent}$$

(c) $f_1' = \dfrac{50}{101} = 0.5 \text{ Hz}$

$f_2' = 30 \text{ kHz} \times 101 = 3.03 \text{ MHz}$

(d)   With feedback $= 500 (30\,000 - 50) = 14.975 \text{ MHz}$

Without feedback $= 4.95 (3\,030\,000 - 0.5) = 15 \text{ MHz}$

(e) $\dfrac{5}{101} = 0.05$ per cent

(f) $A_v\beta = -100$

(g) $1 - A_v\beta = 101$

(h) $20\log\left|\dfrac{1}{1-A_v\beta}\right| = 20\log\dfrac{1}{101} = -40\,\text{dB}$

(i) $\dfrac{500}{4.95} = 101$.

Example 2

A wide-band amplifier has a gain of -1000 without feedback and -20 with negative feedback. Find:

(a) the value of β

(b) the percentage reduction in gain with feedback if the gain without feedback falls by 40 per cent.

Solution

(a) $\qquad A_{vf} = \dfrac{A_v}{1 - A_v\beta}$

Hence

$$\dfrac{-1000}{1 - A_v\beta} = -20$$

$$1 - A_v\beta = \dfrac{1000}{20} = 50$$

$$A_v\beta = 1 - 50 = -49$$

$$\beta = \dfrac{-49}{-1000} = 0.049$$

(b) Gain falls by 40 per cent, i.e. falls to -600

$$1 - A_v\beta = 1 - 0.049\,(-600)$$

$$= 1 + 29.4 = 30.4$$

$$A_{vf} = \dfrac{-600}{30.4} = -19.7$$

$$\text{percentage fall} = \dfrac{20 - 19.7}{20} \times 100 = 1.5 \text{ per cent.}$$

Example 3

An amplifier has a gain of 50 with a $180°$ phase change and a second harmonic distortion of 5 per cent. Negative feedback is used to reduce this distortion to 0·5 per cent. Find the gain of the preamplifier required to maintain the output constant.

Solution

$$\frac{5}{1 - A_v \beta} = 0.5$$

$$1 - A_v \beta = \frac{5}{0.5} = 10$$

Hence as this is the reduction in gain produced in the main amplifier it is also the gain required in the preamplifier.

Example 4

A d.c. amplifier (i.e. one whose frequency response extends down to zero frequency) has a gain of 1000 and a bandwidth of 5 kHz. The gain is reduced by applying −40 dB of feedback. What is the new bandwidth?

Solution

Gain-bandwidth product $= 5000 \times 1000 = 5 \times 10^6$ Hz

$$20 \log \left| \frac{A_{vf}}{1000} \right| = -40$$

$$\log \left| \frac{A_{vf}}{1000} \right| = -2$$

$$\left| \frac{A_{vf}}{1000} \right| = 0.01$$

$$A_{vf} = 10$$

$$\text{Bandwidth} = \frac{5 \times 10^6}{10} = 500 \times 10^3 \text{ or } 500 \text{ kHz.}$$

6.A Test questions

1. Derive the basic feedback equation

$$A_{vf} = \frac{A_v}{1 - A_v \beta}$$

2. $A_v \beta$ is negative for negative feedback.
   (a) true
   (b) false

3. $A_v \beta$ can never be positive.
   (a) true
   (b) false

4. The case of $A_v \beta = 1$
   (a) can never occur
   (b) leads to instability
   (c) means that the feedback is negative
   (d) means that there is no feedback

5. $A_v \beta$ positive and less than one
   (a) can never occur
   (b) leads to instability
   (c) is a case of positive feedback
   (d) is a case of negative feedback

6. A_v is called
   (a) the loop gain
   (b) the gain with feedback
   (c) the feedback factor
   (d) the gain without feedback

7. $1 - A_v \beta$ is called
   [Use the same key as in question 6.]

Questions 8–15 refer to an amplifier with a gain of −10 000 in a negative feedback circuit with a β of 1/50.

8. The gain with feedback is
   (a) 49·75   (c) −50·25
   (b) −50   (d) −49·75

9. The loop gain is
   (a) 201   (c) −10 000
   (b) −200   (d) 1/50

10. The feedback factor is
    (a) 201   (c) −10 000
    (b) −200   (d) 1/50

11. The amount of feedback is
    (a) −46 dB   (c) −34 dB
    (b) 80 dB   (d) −80 dB

12. If the gain of the amplifier (without feedback) drops by 20 per cent, the gain with feedback will fall by
    (a) 20 per cent   (c) 0·4 per cent
    (b) 0·12 per cent   (d) 1 per cent

13. The value of f_1 for the amplifier is 25 Hz. Find its value with feedback (f_1')
    (a) 25 Hz   (c) 5 kHz
    (b) 25 kHz   (d) 0·12 Hz

14. The value of f_2 is 50 kHz. Its value with feedback (f_2') is
    (a) 50·2 kHz   (c) 249 Hz
    (b) 10·1 MHz   (d) 50 kHz

15. The amplifier, without feedback, produces 5 per cent distortion. The percentage distortion with feedback, assuming that the output voltage is kept constant, is
    (a) greater than 5 per cent   (c) 0·05 per cent
    (b) 0·025 per cent   (d) still 5 per cent

16. An amplifier has a gain of -5000 with a tolerance of 35 per cent (i.e. the gain might rise or fall by as much as 35 per cent). If it is to be used in a negative feedback circuit the gain of which is to be constant within 1 per cent find the feedback fraction (β) and the nominal gain with feedback.

17. A three stage amplifier has a gain, in each stage, at normal frequencies, of 100 with a $180°$ phase shift. If negative feedback is applied with a β of $0 \cdot 1$ find the overall gain with feedback.

18. At a frequency of 1 MHz stray capacitance has affected each stage of the amplifier in the previous question such that the output (of the complete amplifier) is now in phase with the input, and the gain has fallen. Determine the value below which the gain must have fallen to ensure that the circuit will not oscillate.

19. An amplifier has a gain of 100 (no phase shift) and positive feedback is applied with a β of $0 \cdot 008$. Find the gain.

20. The gain of the amplifier without feedback of question 19 falls by 50 per cent. Find the percentage fall in gain with feedback.

21. Find the value to which the gain of the amplifier of question 19 must rise for oscillation to commence.

6.3 Practical circuits

Unbypassed emitter resistor

A simple and commonly used method of applying negative feedback is to use an emitter bias resistor without a capacitor in parallel, as shown in Fig. 6.9.

The actual input to the amplifier, v_i, is between the base and emitter as can be seen. This would be equal to v_{in} if a large value capacitor was shorting the emitter to earth, as was seen in Fig. 4.10. The feedback

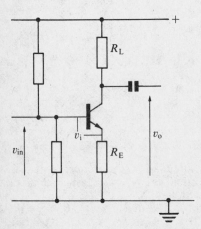

Fig. 6.9. *Unbypassed emitter resistor.*

voltage is that across R_E. The input v_i *sees* v_{in} (base/earth) and the feed-back voltage in series, hence it is an example of *series* feedback, with a high input resistance. In fact, the input resistance is given by

(6.10)
$$R_{in} = h_{ie} + (1 + h_{fe}) R_E$$

as was found for the emitter follower in eqn (5.4).

Many students find difficulty in determining whether the feedback is voltage or current derived. A method of doing this is to imagine that the load is shorted out. This will clearly reduce the output voltage to zero and, if the feedback voltage vanishes at the same time, the feedback must have been proportional to the output *voltage*. An output current will, however, still exist in the short circuit and hence if the effect of shorting out the load does not kill the feedback it must be *current* feedback.

Figure 6.10 shows the effect of shorting out the load in Fig. 6.9. The collector current still flows in R_E so that a feedback voltage still exists and the feedback must be current derived. The circuit of Fig. 6.9 is an example of *current-series feedback*.

A full analysis of this circuit is not possible in this text, but an idea of the approximate voltage gain may be obtained by remembering that V_{BE} is almost constant (0·6 V for silicon). If it were constant then the *alternating* voltage across R_E would equal the alternating input voltage (but 0·6 V d.c. less). Thus the a.c. current in R_E would be v_{in}/R_E and as the same current flows in R_L (apart from the negligible base current) the voltage across R_L, the output voltage, v_o, is given by

$$v_o = \frac{v_{in} R_L}{R_E}$$

(6.11)   Thus $A_{vf} = \dfrac{v_o}{v_{in}} = \dfrac{R_L}{R_E}$ with a 180° phase shift.

This could also have been seen by noting that the ratio of feedback voltage to output voltage is R_L/R_E. This is β so that

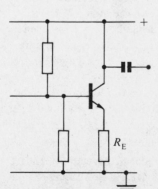

Fig. 6.10. Effect of shorting out R_L *in Fig. 6.9.*

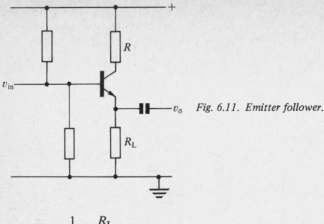

Fig. 6.11. Emitter follower.

$$A_{vf} \simeq \frac{1}{\beta} = \frac{R_L}{R_E}$$

As this is current feedback the output resistance of this circuit will be high.

Omitting the bypass capacitor gives the advantages of reduced distortion, increased bandwidth, etc., as with any negative-feedback circuit, at the expense, of course, of gain.

Emitter follower

A circuit which looks deceptively similar to the last one is the emitter follower, an example of which is shown in Fig. 6.11. The output is now taken from the emitter resistor, labelled R_L. Because the output is in phase with, or follows, v_{in}, the circuit is usually called an emitter follower. Sometimes the resistor R is omitted, as was the case in Fig. 4.13.

The gain of the emitter follower was found using the equivalent circuit in Chapter 4, eqn (4.4). It is a circuit where *all* the output voltage is fed back and hence β is 1 and the gain of the circuit $1/\beta$ is about 1, which agrees with the solution to the problem based on eqn (4.4).

This circuit is an example of voltage feedback because if the load, R_L, is shorted out the feedback vanishes. The input, of course, sees the same input circuit as in Fig. 6.9 and the input resistance is again high and given by eqn (6.10). So the emitter follower has a high input–resistance and a low output–resistance, leading to the type of applications outlined in Chapter 4.

Similar circuits exist using FETs (source followers) and an example can be seen in Fig. 6.12. They too produce extremely high input resistances.

Multistage amplifiers

Fig. 6.13 shows a three-stage amplifier with negative feedback. You might like to decide on the type of feedback before reading on. Like the circuit

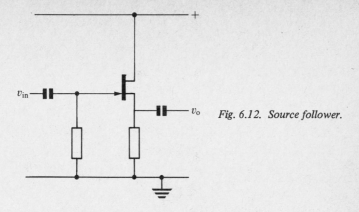

Fig. 6.12. Source follower.

of Fig. 6.9 it is an example of current/series feedback. In fact it is really very similar to that circuit with a gain approaching R_L/R_E. Being a three-stage amplifier it has a higher gain without feedback and will, in fact, be much nearer to R_L/R_E than the amplifier of Fig. 6.9.

Fig. 6.14 shows a two-stage amplifier with feedback. It is negative feedback because when the input goes positive the output also goes positive (two common-emitter stages each producing $180°$ phase shift) causing the fed-back voltage to rise. However a rise on the emitter of T_1 is equivalent to a fall on its base—hence the feedback is negative.

It is voltage feedback because shorting out v_o will kill the feedback, and it is series feedback because the fed-back voltage, across R_1, is in series with the input, v_{in}, as seen by the base-emitter terminals.

The amount fed back is approximately determined by the potential divider R_1 and R_2 and will be

$$\frac{R_1}{R_1 + R_2} = \frac{0 \cdot 15}{5 \cdot 6 + 0 \cdot 15} = \frac{0 \cdot 15}{5 \cdot 75}.$$

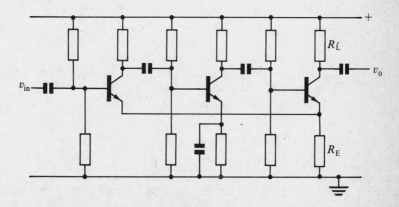

Fig. 6.13. Three-stage feedback amplifier.

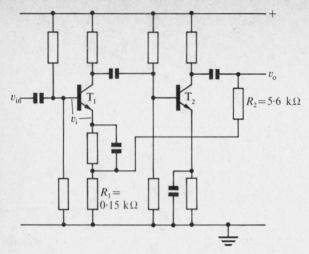

Fig. 6.14. Two-stage feedback amplifier.

As this is β the gain will be about

$$A_{vf} = \frac{1}{\beta} = \frac{5 \cdot 75}{0 \cdot 15} = 38 \cdot 3.$$

Note how the gain is determined by the values of two resistors.

Operational amplifiers

Operational amplifiers have already been mentioned in section 5.6. They are wide-band, high-gain amplifiers, normally with two inputs, one of which produces a 180° phase shift and one with zero phase shift. These are called the inverting and non-inverting inputs respectively. The circuit symbol is shown in Fig. 6.15 with the two inputs marked $-$ and $+$ respectively.

The amplifier, with suitable external components, can be used to perform many mathematical operations, hence the name. The circuit shown in Fig. 6.16 is an example of negative feedback and again you might like to consider what type of feedback it is.

It is, of course, voltage/shunt feedback. An approximate analysis can be made as follows. (It is, in fact, almost exact if A_v is very large.)

With a very large value of A_v, v_i will be very small and can be neglected.

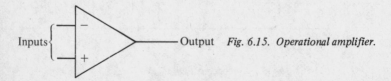

Inputs {   Output   *Fig. 6.15. Operational amplifier.*

This means that point X is effectively at earth potential and it is often called a virtual earth. In this case

$$i_{in} = \frac{v_{in}}{R_1}$$

and $\quad i_f = \dfrac{-v_o}{R_f} \quad$ (minus because i_f flows towards v_o)

Also if v_i is negligible there will be virtually no current flowing into the amplifier input and

$$i_{in} = i_f$$

$$\frac{v_{in}}{R_1} = \frac{-v_o}{R_f}$$

(6.12) $\qquad$ and $\quad A_{vf} = \dfrac{v_o}{v_{in}} = \dfrac{-R_f}{R_1}$

Like other negative–feedback amplifiers the gain does not depend on the amplifier itself. The mathematical operation being performed is multiplication by a constant.

Two points should be noted:

(i) two power supplies (positive and negative) are required to operate the amplifier.

(ii) It is better to return the non-inverting amplifier to earth via a resistor R equal to R_1 and R_f in parallel than to do so directly as in Fig. 6.16

(6.13) $\qquad R = \dfrac{R_1 R_f}{R_1 + R_f}$

More than one input can be used to produce a summing amplifier, an example of which is illustrated in Fig. 6.17.

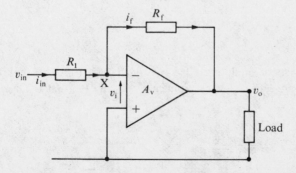

Fig. 6.16. *Multiplication by a constant.*

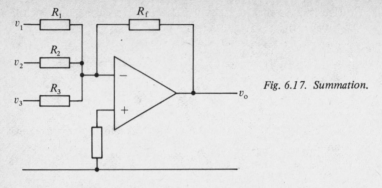

Fig. 6.17. Summation.

If $R_1 = R_2 = R_3 = R_f$

it can easily be shown that

(6.14) $v_o = -(v_1 + v_2 + v_3)$

If the resistors are not equal, v_1, v_2, and v_3 are multiplied by the appropriate resistance ratio and added.

(6.15) $v_o = -R_f \left(\dfrac{v_1}{R_1} + \dfrac{v_2}{R_2} + \dfrac{v_3}{R_3} \right)$

A commonly used operational amplifier is the 741 (integrated circuit). This has a gain of 200 000 and a bandwidth of 10 Hz. This latter is the value of f_2, f_1 being zero. The gain–bandwidth product is thus 2 MHz. The bandwidth is purposely made very low to ensure that the amplifier is stable when feedback is applied. The amplifier will only be used in feedback circuits (such as those of Figs. 6.16 and 6.17) and the gain, with feedback, will be relatively low, ensuring a fairly high bandwidth. For example, if the gain were 100 the bandwidth would be

$$\frac{2 \cdot 10^6}{100} = 2 \cdot 10^4 \, \text{Hz} = 20 \, \text{kHz}$$

At a gain of 10 it would be 200 kHz and so on.

Examples

Example 1

An operational amplifier with a gain of 200 000 (180° phase shift) is used with negative feedback to produce a gain of 10. What is the value of β?

Solution

$$A_{vf} = \frac{A_v}{1 - A_v \beta}$$

$$-10 = \frac{-200\,000}{1 - A_v \beta}$$

$$1 - A_v \beta = 20\,000$$

$$A_v \beta = -19\,999$$

$$\beta = \frac{-19\,999}{-200\,000} = 0 \cdot 1$$

Example 2

If a similar operational amplifier is used in the circuit of the first example but the gain is only 100 000, find the gain with feedback.

Solution

$$A_v \beta = -100\,000 \cdot 0 \cdot 1 = -10\,000$$

$$1 - A_v \beta = 10\,001$$

$$A_{vf} = \frac{-100\,000}{10\,000} = -9 \cdot 99$$

These two examples should convince you that in this type of circuit the gain of the actual amplifier is not very important.

Stabilized power supplies

We have dealt at length in Chapter 3 with power supplies. In many applications, such as d.c. amplifiers, it is very important that the output voltage of the supply remains constant, even if other factors such as the input (mains) voltage or load current vary. Simple stabilizing circuits using zener diodes were discussed.

Negative feedback can be used in stabilizing circuits. Fig. 6.18 shows a block diagram of a series stabilizer. The output is sampled and the sample compared with a reference standard. Any difference is amplified and applied to the control device in such a sense as to change the output in the required direction (negative feedback). A simple, and frequently used circuit, is given in Fig. 6.19.

The fraction of the output is selected by the potential divider R_1 and R_2 and the difference between this and the reference voltage provided by

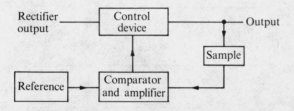

Fig. 6.18. Block diagram of a feedback voltage stabilizer.

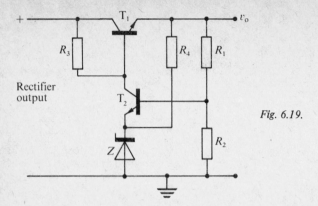

Rectifier
output

Fig. 6.19.

the zener diode (Z) is applied to the transistor T_2 which acts as an amplifier with a load R_3. The output of T_2 is applied to the base of the control transistor T_1. T_1 acts as an emitter follower, v_o being about 0·6 V less than T_1 base.

Hence if v_o, say, rises for any reason, the voltage on the base of T_2 rises, the current in T_2 rises, and its collector voltage, which is also the base voltage of T_1, falls, causing the emitter voltage of T_1, v_o, to fall. R_4 is used to provide a suitable current through the zener diode.

Numerous variations on this circuit are possible, leading to highly sophisticated designs (T_2 might, for example, be replaced by a high-gain operational amplifier).

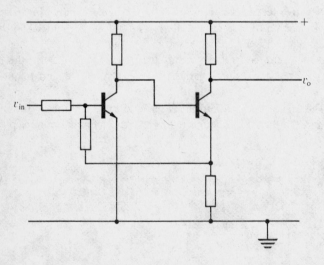

Fig. 6.20. *Questions 1–4 (6.B).*

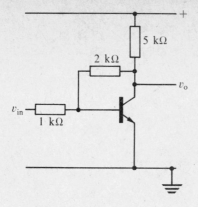

Fig. 6.21. Questions 5–7 (6.B).

6.B Test questions

1. Fig. 6.20 shows a two-transistor amplifier with some biasing components missing. Is this voltage- or current-feedback?

2. Is the circuit of Fig. 6.20 series or shunt feedback?

3. Will the input resistance be high or low?

4. Will the output resistance be high or low?

5. Fig. 6.21 shows a negative feedback circuit. Is it voltage- or current-feedback?

6. Is this feedback series or shunt?

7. The approximate voltage gain of the circuit of Fig. 6.21 is
    (a) 2   (c) 5
    (b) 2·5   (d) 0·4
(Hint—it is similar to an operational amplifier.)

8. Fig. 6.22 shows a circuit with two outputs. What is the approximate voltage gain from the input to each output?

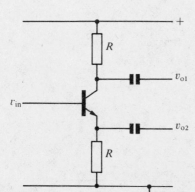

Fig. 6.22. Questions 8–11 (6.B).

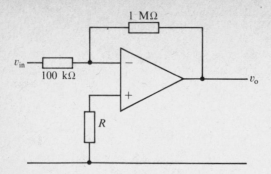

Fig. 6.23. *Questions 12–13 (6.B)*.

9. Is this feedback voltage or current? (Consider each output separately.)

10. Is the output resistance high or low? (Again consider each output separately.)

11. If the two outputs were viewed on a double-beam oscilloscope what difference would be observed?

12. In Fig. 6.23 $v_{in} = 1 \, V$. v_o will be
    (a) −1 V   (c) 10 V
    (b) −10 V   (d) −1/10 V

13. The optimum value for R is
    (a) 1 MΩ   (c) 100 kΩ
    (b) 1·1 MΩ   (d) 91 kΩ

14. Fig. 6.24 shows a summing amplifier. Calculate the output voltage.

15. Find the output voltage in the operational amplifier of Fig. 6.25.

16. An operational amplifier has an open circuit gain of 100 dB and a value of f_2 of 5 Hz. It is used in the circuit of Fig. 6.26. Find the bandwidth.

17. What is the maximum gain which the amplifier of the previous question could give if the bandwidth is to be at least 1 MHz?

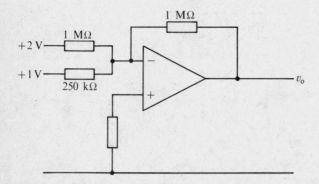

Fig. 6.24. *Question 14 (6.B)*.

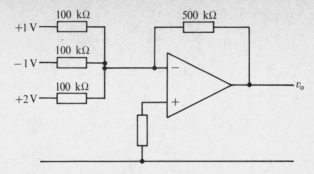

Fig. 6.25. Question 15 (6.B).

18. Draw a block diagram of a negative feedback voltage stabilizer.

19. Sketch a circuit diagram of a feedback voltage stabilizer employing a control transistor, an amplifying transistor, and a zener diode. Explains its operation. Suggest a method which might be used to produce a variable ouput voltage.

20. Fig. 6.27 shows a current/series negative feedback amplifier. It employs a transistor with an h_{fe} of 200 and an h_{ie} of $1 \cdot 5$ kΩ. Find the input resistance and the approximate voltage gain, assuming that the values of R_1 and R_2 are very high.

21. The gain of the amplifier of Fig. 6.27 without feedback (i.e. with a large value capacitor across R_E) is given by

$$A_v = \frac{-h_{fe} R_L}{h_{ie}}$$

Use this formula and a value of β of R_E/R_L to find a more accurate estimate of the gain with feedback.

22. Calculate the input resistance to Fig. 6.27 using the more realistic values of R_1 and R_2 of 84 kΩ and 16 kΩ respectively.

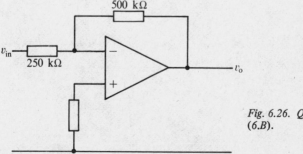

Fig. 6.26. Questions 16–17 (6.B).

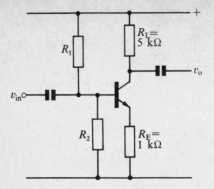

Fig. 6.27. Questions 20–22 (6.B).

7 Waveform generation and shaping

7.1 Introduction

We are all very familiar with the *sine wave*. It is the fundamental waveform of engineering, and is the waveform produced when a coil rotates in a uniform magnetic field. It is, of course, the waveform used by the generating authorities who supply homes and industry with power.

Sinusoidal waveforms have a particularly useful property which is utilized by electrical engineers. If a sinusoidal current is passed through a perfect (that is a linear) resistor, capacitor, or inductor, then the voltage developed across the component is also sinusoidal. For any other waveform the shape of the voltage will differ, perhaps dramatically, from the shape of the current flowing, if there is inductance or capacitance in the circuit.

In electronic engineering we often use sine waves to test circuits such as amplifiers. This gives us much useful information even if the amplifiers are going to be used for amplifying waveforms which are not sinusoidal. A knowledge of how a circuit behaves at various frequencies will enable us to predict how it will respond to a complex input signal. Hence we need to have available sine-wave generators which should be capable of supplying sine waves over a range of frequencies, and with a variable amplitude. As the power required will not be very large, electronic means of generation are used and we shall deal with this subject in section 7.3.

Many other waveforms are used in electronics, such as pulses, sawtooth waves, and triangular waves. Pulses are used extensively in electronics, for example in digital electronics (see Chapter 8) and, because of the fairly precise time at which their voltage levels change, in many timing applications. In a radar system, for example, a pulse is transmitted and the radar receiver picks up any reflections from objects such as ships or aircraft. The time between transmission and reception of the pulse is a measure of the distance away of the object.

A perfect pulse is shown in Fig. 7.1; a real pulse will have various defects. Firstly there is the fact that voltage levels cannot change instantaneously (this could only happen if inductance and capacitance were

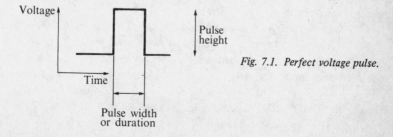

Fig. 7.1. Perfect voltage pulse.

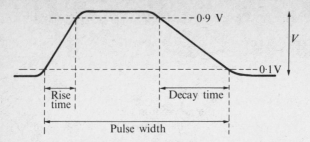

Fig. 7.2. *Pulse, showing rise time, decay time, and width.*

completely absent from the circuit—an impossible situation). This imperfection is measured by the *rise-time* and *decay-time*, defined as the time for the voltage to rise, or fall, from 10 per cent to 90 per cent (or 90 to 10 per cent) of its final level. These are shown in Fig. 7.2, as is the width of the pulse, usually defined in terms of its 10 per cent values (although it is sometimes quoted at the 50 per cent levels).

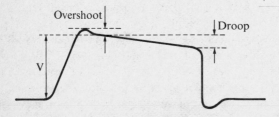

Fig. 7.3. *Pulse, showing overshoot and droop.*

Fig. 7.3 shows two other pulse imperfections, *droop*, often quoted as a percentage of the pulse height, and *overshoot*, which occurs in some circumstances, and which is again quoted as a percentage of the pulse height.

All these imperfections are related to the sinusoidal response of the circuit. The performance of an amplifier is sometimes measured with pulse, or square-wave inputs as well as, or instead of, sinusoidal inputs, and pulse generators are an important laboratory instrument.

Pulse generators usually produce a stream of pulses at regular intervals. This will be necessary not only for the laboratory instrument but also for the radar pulse generator. Two more quantities need to be considered: the number of pulses per second, which is called the *pulse repetition frequency* (p.r.f.), and the ratio of the width of the pulse to the time between pulses. This latter is called the *mark-to-space ratio*, the name coming from early morse-code transmission systems where the pulse was used to cause a pen to *mark* the paper. Fig. 7.4 shows some (perfect) pulses with a mark to space ratio of 1 to 4. The p.r.f. of the waveform can be found because the period of the waveform is $5\,\mu s$, so that there are $1/(5 \times 10^{-6})$ or $200\,000$ pulses per second.

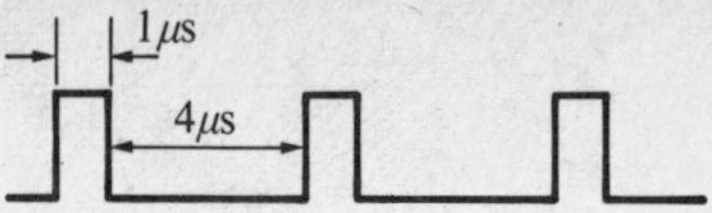

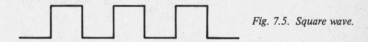

Fig. 7.4. Pulse train.

Another commonly used waveform is the square wave, Fig. 7.5, which is, in effect, a pulse waveform with a 1-to-1 mark-to-space ratio. The square wave will, of course, have a rise time, a decay time, droop, and, perhaps, overshoot. A number of oscillators are available on the market having square wave and pulse outputs as well as a sinusoidal output. One method of generating square waves and pulses will be dealt with in Section 7.4.

Fig. 7.5. Square wave.

One last waveform to be mentioned here is the triangular wave, not to be confused with the saw tooth. Fig. 7.6 shows such a waveform.

Fig. 7.6. Triangular wave.

7.2. Waveform shaping

The previous section referred to the fact that sinusoidal waveforms remain unaltered in shape when passed through linear circuits. This section gives an introduction to the effects of linear circuits on sine waves of different frequencies and on non-sinusoidal waveforms.

R–C (*integrating*) *circuit*

This very simple circuit is shown in Fig. 7.7. We shall consider it from two different viewpoints. Firstly its effect on sine waves of different frequencies. At low frequencies the reactance of C is very large and

(7.1) $$v_{\text{o}} \simeq v_{\text{in}}$$

However, as the frequency is raised the effect of C becomes more important, R and C form a potential divider and the output voltage falls. At high frequencies the output is inversely proportional to frequency, halving for a doubling of frequency and becoming zero at an infinite frequency! As a halving of the output voltage is a fall of 6 dB and a doubling of frequency is a rise of an octave, the response of the circuit at high frequencies is said to be a fall at a rate of 6 dB per octave. Note that the terms high and low as used above are purely relative and that 1 kHz, for example, may be high in one context but low in another.

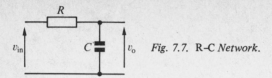

Fig. 7.7. R–C Network.

One other important point about the sinusoidal response of this circuit is that at a frequency such that the reactance of C is equal to R the output will be 3 dB down. This frequency is called f_2 to conform to the definition of bandwidth given in Chapter 5. The circuit is, in fact, a very simple *low-pass filter*, letting through frequencies up to f_2. (This does not mean that all frequencies above f_2 are stopped but only that f_2 is the universally accepted definition of the bandwidth.) As the reactance of C is $1/2\pi fC$ this frequency is given by

(7.2)
$$f_2 = \frac{1}{2\pi RC}.$$

The sinusoidal response of the circuit is shown in Fig. 7.8.

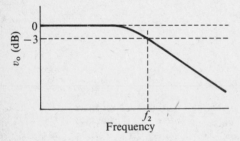

Fig. 7.8. *Sinusoidal response of Fig. 7.7.*

We have already met this circuit in Chapter 3, when a low-pass filter was used for smoothing the output of a rectifier. Sometimes the filter is in circuit accidentally, as will happen in every amplifier where R is the output resistance of the amplifier and C the *stray capacitance* across the output terminals. This explains why, as mentioned in Chapter 5, there is always an upper limit to the frequency of operation of an amplifier.

The output of the circuit will not only vary in amplitude with frequency, but will also vary in phase. At low frequencies, where C has no effect, the input and output are in phase. At f_2, v_o will lag behind v_{in} by 45° and this phase angle will get near to 90° at very high frequencies. Theoretically it will lag by exactly 90° at an infinite frequency, when the output voltage will have fallen to zero!

The second way in which we shall consider the R–C circuit is in its effect on pulses or square waves. Consider a step voltage input (that is a sudden change in input voltage) applied to an R–C circuit. The voltage across the capacitor (v_o in Fig. 7.7) rises exponentially as in Fig. 7.9. It rises to 63·2 per cent (about $\frac{2}{3}$) of its final value in a time equal to CR,

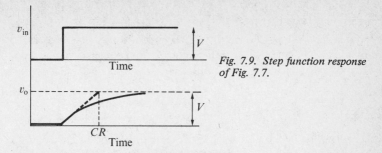

Fig. 7.9. Step function response of Fig. 7.7.

called the *time constant* of the circuit. A pulse is simply a rising step input followed by a falling one, and the effect of the circuit of Fig. 7.7 on the pulse depends on the length of the pulse relative to the time constant.

Figure 7.10 shows the input pulse (a) and three possible outputs.

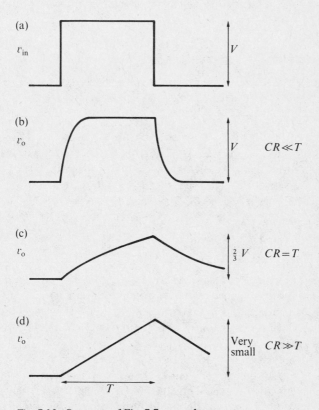

Fig. 7.10. Response of Fig. 7.7 to a pulse.

(b) shows the effect of a *CR* value very much less than the pulse width. The time for the charging of *C* is hardly noticeable relative to the pulse width, and the output has almost the same shape as the input. (c) shows the situation when the CR value is about equal to the pulse width. The output only reaches about 2/3 of V when the pulse finishes, leading to severe distortion. Lastly in (d) we see the response of a circuit with a CR value much longer than the pulse width. The capacitor has only just started to charge up when the pulse ends and an almost linear rise and fall are produced. A set of pulses of a 1-to-1 mark-to-space ratio (i.e. a square wave) will give a triangular wave as output. Mathematically, the output of the circuit (with a large time constant) is approximately the integral of the input and hence the circuit is called an *integrating circuit*. Note that the voltages in Fig. 7.10 are not drawn to the same scale.

C–R (*differentiating*) *circuit*

Figure 7.11 shows a *C–R* circuit. It has occurred as the coupling between stages in many of the amplifiers considered in earlier chapters. It is clearly very similar to the integrating circuit—in fact it is the same circuit with the output across R instead of C.

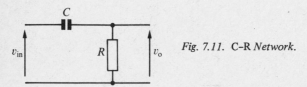

Fig. 7.11. C–R *Network.*

In this case v_o will be greatest at high frequencies, and at very high frequencies where the reactance of C can be neglected, it passes the signal without attenuation. At low frequencies the output falls, becoming zero at zero frequency. The fall is now 6 dB per octave at low frequencies and again when the reactance of C is equal to R it will have fallen by 3 dB. This frequency is called f_1 and the circuit is a simple *high-pass filter*.

(7.3)
$$f_1 = \frac{1}{2\pi RC}.$$

The phase angle is now such that v_o leads v_{in} by 45° at f_1 and becoming 90° lead at zero frequency (with zero amplitude output).

The amplitude response is shown in Fig. 7.12. This response, together with that of Fig. 7.8, explains why the response of an a.c. coupled amplifier is as shown in Fig. 5.13 (p.103).

If a step voltage V is applied as input, the output, v_o, is now a voltage proportional to the charging current. This, you may remember, starts at a value V/R and falls exponentially, Fig. 7.13. Fig. 7.14 (a) shows a pulse applied to this circuit. (b) shows the output if RC is very long relative to the pulse width. The fall in current has only just started when the pulse ends, and the output is almost identical to the input. In (c) the value of RC is about equal to the pulse width, the circuit distorting the pulse considerably. Lastly (d) is the case of an RC value much shorter than the pulse length and the resultant short pulses of charge and discharge current

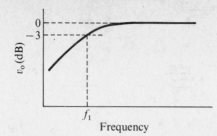

Fig. 7.12. Sinusoidal response
of Fig. 7.11.

produce an output voltage of the shape shown. It bears no resemblance to
the input, and is, in fact, approximately the mathematical derivative of
v_{in}. The circuit is usually called a *differentiating circuit*. Of course, if the
C-R circuit is part of an amplifier which is meant to amplify the pulse,
CR must be made as large as possible to avoid this differentiation. The
circuit, with a low CR, is often used to produce short pulses from, for
example, a square wave.

7.A Test questions

1. If a sinusoidal current is passed through a non-linear component the
voltage across the component is also sinusoidal.
    (a) true
    (b) false

2. If a sinusoidal current is passed through a linear resistor the voltage
across the resistor is also sinusoidal.
    (a) true
    (b) false

3. If a sinusoidal current is passed through a linear inductor (or
capacitor) the voltage across the inductor (or capacitor) is also sinusoidal.
    (a) true
    (b) false

4. Discuss the imperfections possessed by a real pulse relative to a
perfect one.

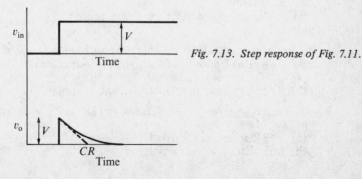

Fig. 7.13. Step response of Fig. 7.11.

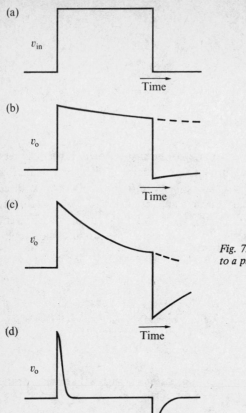

(a)

v_{in}

Time

(b)

v_o

Time

(c)

v_o

Time

Fig. 7.14. Response of Fig. 7.11 to a pulse.

(d)

v_o

T

Time

5. Define p.r.f.

6. Define mark-to-space ratio.

7. Define the terms
    (i) rise time,   (ii) decay time,   (iii) width
of a pulse.

8. A train of pulses has a p.r.f. of 50 000 pulses per second and the pulses have a width (duration) of $2\,\mu s$. The mark-to-space ratio is
    (a) $1:10$ (c) $1:25\,000$
    (b) $1:9$ (d) there is insufficient information

9. An integrating circuit, Fig. 7.7, has $R = 1\,M\Omega$ and $C = 100\,pF$. The value of f_2 will be
    (a) $1.59\,kHz$ (c) $10\,kHz$
    (b) $1.59\,MHz$ (d) infinity

10. At f_2 the reactance of C will be
 (a) zero   (c) 100 kΩ
 (b) 1 MΩ   (d) 1·59 MΩ

11. At this frequency the impedance of the circuit will be
 (a) 2 MΩ   (c) 1·414 MΩ
 (b) 1 MΩ   (d) very low

12. If a sine wave of frequency f_2 and amplitude 1 V is applied, the amplitude of the current flowing will be
 (a) 0·707 μA   (c) 1 μA
 (b) 0·707 mA   (d) 1·414 μA

13. This current will give a voltage across C (v_o) of amplitude
 (a) 1·414 V   (c) 0·707 V
 (b) 0·707 μV   (d) 1 V

14. This voltage, relative to the input, is
 (a) +3 dB   (c) +6 dB
 (b) −3 dB   (d) −6 dB

15. The phase angle of the current (question 13) relative to the input voltage is
 (a) zero   (c) 45° lag
 (b) 45° lead   (d) 90° lead

16. The phase of v_o relative to the current is
 (a) zero   (c) 90° lag
 (b) 90° lead   (d) 45° lead

17. The phase angle of v_o relative to v_{in} is
 (a) 45° lead   (c) 90° lag
 (b) 45° lag   (d) zero

18. At 1 MHz the output of this circuit for a certain input is 10 V. Estimate the output at 2 MHz
 (a) 10 V   (c) 4 V
 (b) 7·07 V   (d) 5 V

19. At 10 Hz the output of this circuit for a certain input is 10 V. Estimate the output at 5 Hz.
 (a) 10 V   (c) 4 V
 (b) 7·07 V   (d) 5 V

20. Three pulses of widths 1 μs, 100 μs, and 10 ms respectively are applied to the circuit of question 9. The pulse which will be most nearly integrated will be the one of width
 (a) 1 μs   (c) 10 ms
 (b) 100 μs

21. The time constant of this circuit is
 (a) 1 μs   (c) 10 ms
 (b) 100 μs   (d) 1 s

22. If the output is now taken across R instead of C, differentiation will be most nearly perfect (considering the three pulses of question 20) for the pulse of width
 (a) 1 μs   (c) 10 ms
 (b) 100 μs

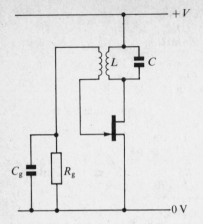

Fig. 7.15. *Tuned drain oscillator.*

7.3 Sinusoidal oscillators

We have seen in the previous chapter that an amplifier will oscillate if $A_v\beta = 1$ where A_v is the gain without feedback and β is the fraction of the output which is fed back. In this case the amplifier just provides enough gain to overcome the losses in the feedback network. The output just provides sufficient signal via the feedback network to maintain itself.

A sinusoidal oscillator must be arranged so that this condition occurs at only one frequency, and hence the feedback network must be frequency selective, i.e. it must contain inductance and/or capacitance. Two basic types of sinusoidal oscillator will be considered, one whose feedback path contains inductance and capacitance (the *L–C* oscillator) and one which contains capacitance and resistance (the *R–C* oscillator). An *L–R* oscillator is, of course, possible but would no advantages over an *R–C* circuit and would have the disadvantage of the size and cost of the inductor.

L–C *oscillators*

A very simple oscillator is the tuned collector, or tuned drain circuit. This is similar to the r.f. amplifier of Fig. 5.17 (p.108) with the input coupled back to the output, Figs. 7.15 and 7.16. As the amplifier produces 180° phase shift the transformer must be wound to produce 180° shift. This circuit will oscillate as long as the value of g_m or h_{fe} is sufficient. The minimum value can be calculated in terms of the other circuit parameters, including M, the mutual inductance between the transformer windings. The frequency of oscillation of this, and most other LC oscillators, is given approximately by

(7.4) $$f = \frac{1}{2\pi\sqrt{(LC)}}$$

which is about the resonant frequency of the tuned circuit. This formula will be fairly accurate as long as the resistance of the coil is small.

Unfortunately, if $A_v\beta$ is larger than 1 the oscillations will increase in amplitude until severe distortion occurs. It is clearly not possible to design a satisfactory circuit in which $A_v\beta$ is *exactly* 1 as any slight change in one of the components due to temperature, etc., would alter the value of $A_v\beta$. Both circuits shown include a method of bias which automatically maintains $A_v\beta$ at 1.

Circuits similar to Figs. 7.15 and 7.16 can be designed with the tuned circuit in the gate or base circuit instead of the drain or collector.

Mention should be made of two very importance L–C oscillators, the *Hartley* and the *Colpitts*. These exist in many forms and an example of each is shown in Figs. 7.17 and 7.18 respectively. The frequency of oscillation of the Hartley oscillator is given by

$$(7.5) \qquad f = \frac{1}{2\pi\sqrt{LC}},$$

where L is the series combination of L_1 and L_2:

$$(7.6) \qquad L = L_1 + L_2 \quad \text{(assuming no mutual coupling).}$$

For the Colpitts oscillator the frequency is also given by eqn 7.5 where C is now the parallel combination of C_1 and C_2

$$(7.7) \qquad C = \frac{C_1 C_2}{C_1 + C_2}.$$

In both cases, of course, the gain must be sufficient to maintain oscillation. It can be shown that for a transistor oscillator h_{fe} must be greater than L_2/L_1 in the case of the Hartley and C_1/C_2 for the Colpitts. However in an FET circuit we require μ greater than L_1/L_2 or C_2/C_1 for the Hartley and Colpitts respectively.

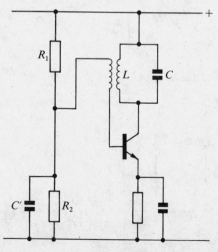

Fig. 7.16. *Tuned collector oscillator.*

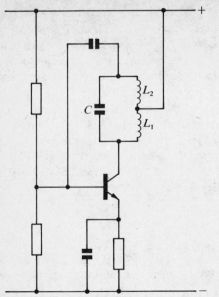

Fig. 7.17. Hartley oscillator.

R–C oscillators

In an *R–C* oscillator the frequency-determining part of the feedback network consists of resistance and capacitance. The first type that we shall consider is the *phase-shift oscillator*.

The name of this circuit comes from the fact that a single transistor is used, producing $180°$ phase shift, together with a network of *R*s and *C*s to provide another $180°$ shift at one frequency. Now, as we have seen in section 7.2, a single *R–C* network of the type shown in Fig. 7.19 will

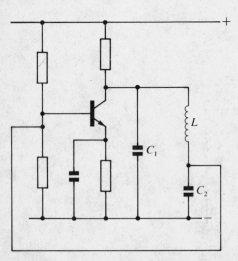

Fig. 7.18. Collpitts oscillator.

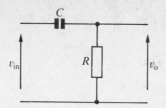

Fig. 7.19. C–R *Network.*

produce an output v_o which leads the input v_{in}. However, the greatest lead that can be produced is $90°$ and this will only occur in circumstances which would mean that the magnitude of v_o is zero!

Two such networks cascaded could give a $180°$ shift, but again with zero output. However, three networks, each producing a $60°$ phase shift would produce a total of $180°$ and give a finite output voltage. In fact, they do not need to produce $60°$ each as long as the sum of the three shifts is $180°$. The derivation of the frequency of oscillation is quite complex, because each network affects the previous one, but if the three Cs and the three Rs are equal (a convenient, but not essential requirement), it can be shown that a $180°$ shift is produced at a frequency given by

(7.8) $$f = \frac{1}{2\pi RC \sqrt{6}}.$$

It is also found that such a circuit, Fig. 7.20, produces a loss, at this frequency, of 29, i.e.

(7.9) $$v_o = \frac{v_{in}}{29}.$$

Hence the amplifier must have a gain of at least 29. An FET phase-shift oscillator is shown in Fig. 7.21.

A transistor circuit would be similar, although the low input resistance of the transistor affects the result of eqn (7.9) and, in fact, the lowest possible value of h_{fe} which can be used is 44·5.

The other type of R–C oscillator to be considered is known as a *Wien bridge oscillator.* (The name derives from a type of bridge circuit used in measurements, the feedback network of the oscillator being part of a balanced Wien bridge.) This oscillator uses an amplifier with zero phase

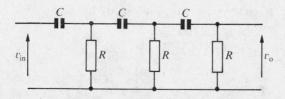

Fig. 7.20.

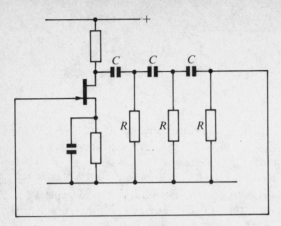

Fig. 7.21. Phase-shift oscillator.

shift, for example by using two transistors, and the feedback network is a circuit which will give no phase shift at the frequency of oscillation.

An example of the type of network used is shown in Fig. 7.22. It can be shown that the network gives zero phase shift at a frequency given by

$$(7.10) \qquad f = \frac{1}{2\pi \sqrt{(R_1 R_2 C_1 C_2)}} \, .$$

However, if $R_1 = R_2 = R$ and $C_1 = C_2 = C$, which is often the case, eqn (7.10) reduces to

$$(7.11) \qquad f = \frac{1}{2\pi RC} \, .$$

At this frequency the attenuation produced if the Rs and Cs are equal is three. A transistor Wien bridge oscillator is shown in Fig. 7.23. The transistor amplifier $T_1 T_2$ provides the gain (zero phase shift) and the frequency-determining network is shown (two Rs and two Cs). Of course, the two-stage amplifier would give far too much gain, and negative feedback is provided via R_T. This is a *thermistor*, a device with a negative temperature coefficient of resistance. When first switched on R_T has a high resistance and the amount of negative feedback applied is small. As the oscillations build up, the temperature of R_T rises, it reduces in value and more negative feedback is applied, reducing the gain. The oscillations will stabilize when the amplifier gain is exactly three.

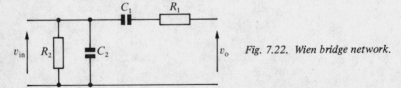

Fig. 7.22. Wien bridge network.

R–C oscillators can be made variable in frequency by using variable components for the Rs or Cs. Variable resistors are normally cheaper and a ganged component is usually used.

In both phase shift and Wien bridge oscillators the frequency is inversely proportional to RC, whereas in an $L–C$ oscillator it is inversely proportional to $\sqrt{LC}$. At very low frequencies where RC or LC needs to be very large, there are clearly advantages in the $R–C$ type. Firstly a halving of frequency will double R or C but in an $L–C$ type it will increase L or C four times, meaning that low-frequency oscillators would need very large values of L or C and these components will be physically large and will also be expensive. Low-frequency $R–C$ oscillators will require smaller values of C and can, of course, employ large values of R, the latter not being physically very large or expensive.

Consider a 1 Hz oscillator using, say, a 1 μF capacitor. An $L–C$ oscillator will need an inductor of value

$$L = \frac{1}{4\pi^2 f^2 C}$$

$$= \frac{1}{4\pi^2 (1)^2 (10^{-6})^2}$$

$$= \frac{10^{12}}{4\pi^2} = 2\cdot 5 \times 10^{10} \text{H},$$

a ridiculously large value.

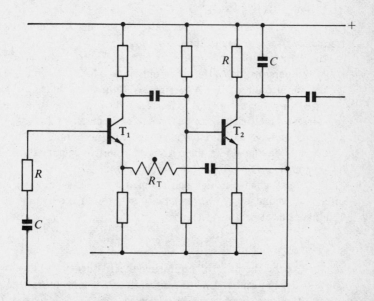

Fig. 7.23. Wien bridge oscillator.

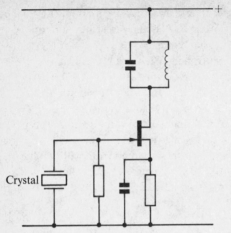

Fig. 7.24. Crystal oscillator.

However an R–C phase-shift oscillator with a 1-μF capacitor requires a resistor of value

$$R = \frac{1}{2\pi\,Cf\,\sqrt{6}}$$

$$\frac{10^6}{15\cdot4}\,\Omega = 65\ \text{k}\Omega,$$

a very reasonable value.

Factors affecting frequency stability

The *frequency stability* of an oscillator is its ability to maintain a constant frequency over a period of time.

The most obvious and important factor affecting frequency is temperature. Temperature will affect not only the values of R, L, and C in the circuit but also the transistor parameters. Compensation can be used and is effective to a limited extent. Thus capacitors with positive and negative temperature coefficients can be connected in parallel to offset each other.

Supply voltage variations will also affect the frequency of oscillation, due mainly to the effect they have on transistor parameters. (They will have a much smaller effect in RC oscillators.)

Lastly any change in the load coupled to the oscillator will have some effect on frequency. A *buffer amplifier* between oscillator and load can eliminate this effect, however.

Crystal oscillators

As well as the methods of improving the frequency stability mentioned above, a crystal can be used to control the frequency. Most of us will have heard of *piezoelectric crystals* as devices used in some record players.

A force on the faces of the crystal, caused by the movements of the stylus, produces a potential across the crystal. Similarly, if a potential is applied across the faces of a piezoelectric crystal, forces are produced leading to deformations of the crystal. This mechanical system will, like any other, have a resonant frequency and a Q factor. Crystals with resonant frequencies from a few kilohertz to many megahertz with Qs ranging up to several hundreds of thousand are available, with properties which are extremely stable with respect to time and temperature changes. Electrically the crystal acts as a parallel tuned circuit and can be used as such in an oscillator, producing excellent frequency stability. Such a circuit is shown in Fig. 7.24, using an FET, a 'normal' tuned circuit as drain load, and a crystal in the gate circuit.

7.B Test questions

1. Discuss the requirements of an oscillator in terms of its being a feedback amplifier.

2. Draw a circuit of and explain the action of
    (i) a tuned anode
    (ii) a tuned collector oscillator.
Explain the biasing arrangements in each case.

3. Explain what is meant by squegging.

4. The Hartley oscillator shown in Fig. 7.17 has $L_1 = 1\ \mu H$ and $L_2 = 40\ \mu H$.
If $C = 100\ pF$ determine the frequency of oscillation, assuming no coupling between the coils. What is the minimum value of h_{fe} needed to ensure oscillation?

5. A Colpitts oscillator, Fig. 7.18, has $C_2 = 50\ pF$ and is required to oscillate at 1 MHz. If it is essential that it can be used with a transistor of $h_{fe} = 30$, find the required values of C_1 and L.

6. If the Cs and Rs in Fig. 7.20 are interchanged v_o lags behind v_{in} instead of leading it. Explain why it is still possible to use this network in a phase shift oscillator.

7. Explain why the amplifier used in the phase shift oscillator of Fig. 7.21 must have a minimum gain of 29.

8. Would the circuit of Fig. 7.21 still oscillate if the three Rs and Cs had different values, and if so would the minimum gain still be 29?

9. Two R–C networks can give a phase shift of $180°$. Why can such a network not be used in the feedback circuit of the phase shift oscillator?

10. A phase shift oscillator of the type shown in Fig. 7.21 had a three-gang variable capacitor ranging from $0·01\ \mu F$ to $1\ \mu F$ and a value of R of $2\ k\Omega$. Find the frequency range of the oscillator.

11. The same capacitor is used in a tuned collector L–C oscillator. If the highest frequency is to be the same as in Question 10 determine the required value of L.

12. In Question 11 what range of frequencies is covered by the oscillator?

13. A Wien bridge oscillator, as in Fig. 7.23 is designed to cover the range of frequencies 1 — 1000 Hz. If $C = 1\ \mu F$, what range of resistance values is required?

14. Discuss the relative merits of L–C and R–C oscillators at very low frequencies.

15. Discuss the factors which govern the frequency stability of an oscillator.

16. Explain, in simple terms, how a piezoelectric crystal can be used in the design of a very stable oscillator, including a circuit diagram of such a circuit.

7.4 Multivibrators

There are three basic types of multivibrator: astable, monostable, and bistable. Only one of these, the astable, oscillates, or vibrates. Some authorities argue that the others should not be called multivibrators, but they are so widely known as such, that the name will be used here.

All three consist of two transistors, valves, or FETs each coupled to the other as shown in Fig. 7.25. The three types differ in the type of coupling networks employed. The transistors are not being used as amplifiers in the normal way, and the circuits are such that either T_1 conducts heavily (output 1 at almost zero potential) and T_2 is cut off (output 2 at $+V$) or vice-versa. They do not operate in the centres of the load lines (only at either end) and no bias stabilizing networks are required.

The astable multivibrator is not stable, but switches very rapidly from one state to the other, remains there for a length of time depending on the circuit components, and then switches back. It thus oscillates, producing pulses of a certain mark-to-space ratio (shown in Fig. 7.26 with

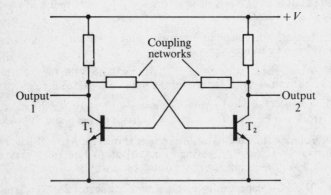

Fig. 7.25. Multivibrator.

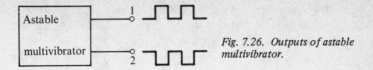

Fig. 7.26. *Outputs of astable multivibrator.*

a one-to-one ratio—i.e. a square wave). As can be seen, two outputs, 180°
out of phase with each other, are available.

The monostable multivibrator is stable in one of its two states, and
remains there until an external signal (usually a pulse) switches it to the
other state where it remains for a predetermined time before switching
back to the stable state (Fig. 7.27).

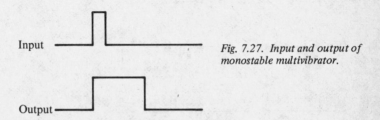

Fig. 7.27. *Input and output of monostable multivibrator.*

The bistable multivibrator is stable in either of its two states, remaining
in that state until an external signal switches it to the other (Fig. 7.28).

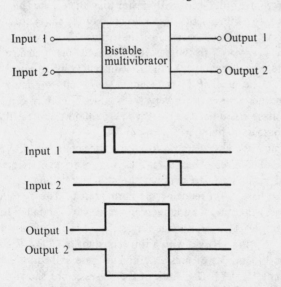

Fig. 7.28. *Inputs and outputs of a bistable multivibrator.*

Astable multivibrator

The understanding of any of the multivibrators will be made much easier if it is remembered that any change of voltage applied to one plate of a capacitor is immediately seen on the other plate, i.e. the voltage across the capacitor cannot change instantaneously.

An astable multivibrator is shown in Fig. 7.29. Note that it is not essential to draw the coupling leads at 45° to the vertical, as shown, but it is usually done and identifies the circuit immediately as a multivibrator.

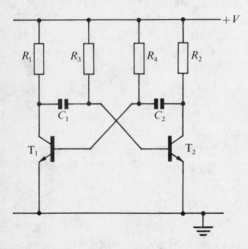

Fig. 7.29. An astable multivibrator.

When the circuit is switched on, one of the transistors will start conducting before the other, or slightly faster than the other. This will initiate a very rapid switching action as follows. Say T_1 starts to conduct. Its collector voltage will fall because of the voltage drop in R_1, and this fall will at once pass via C_1 to the base of T_2 (remember a change of voltage on one plate is accompanied, in the first instance, by an equal change on the other). If the base of T_2 were about zero volts to start with it will now become negative and in a very short time T_1 is fully conducting, its collector voltage has fallen from $+V$ to zero (almost) so that there will be $-V$ on the base of T_2 holding T_2 well cut off.

C_1 now starts to charge up via R_3 with a time constant of $C_1 R_3$ and the base voltage of T_2 rises. When it reaches zero, or to be more precise about 0·6 V positive, T_2 starts to conduct, its collector falling and this fall is passed to the base of T_1 reducing the current in T_1. The T_1 collector (and hence T_2 base) rises and T_2 conducts more heavily, the fall in its collector voltage passing to T_1 base, so that very soon T_1 is cut off and T_2 conducting heavily. C_2 now charges with a time constant of $C_2 R_4$. The output, of either T_1 or T_2, consists of pulses, although these will form a square wave if $C_1 = C_2$ and $R_3 = R_4$. The width of the pulses can be shown to be about 0·7 times the time constant. Hence in the circuit shown the widths are $0·7 C_1 R_3$ and $0·7 C_2 R_4$ respectively. The periodic time of the waveform will be given by

(7.10) $$T = 0·7\,(C_1 R_3 + C_2 R_4).$$

Obviously if $R_3 = R_4 = R$ and $C_1 = C_2 = C$ the frequency of the square wave produced will be

(7.11)
$$f = \frac{1}{1 \cdot 4\,RC}.$$

Typical waveforms at various points are shown in Fig. 7.30.

Example

Problem

In Fig. 7.29, $R_3 = 20\,\text{k}\Omega$, $R_4 = 40\,\text{k}\Omega$, and $C_1 = C_2 = 220\,\text{pF}$. Determine the frequency and mark-to-space ratio of the waveform.

Solution

The times of each part of the waveform are

$$0 \cdot 7 R_3 C_1 = 0 \cdot 7 \times 20 \times 10^3 \times 220 \times 10^{-12}\,\text{s}$$

$$= 3 \cdot 08\ \mu\text{s}$$

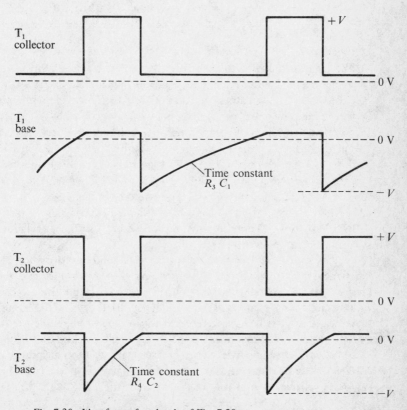

Fig. 7.30. *Waveforms for circuit of Fig. 7.29.*

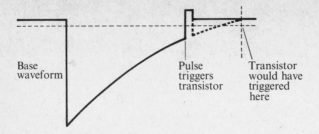

Base waveform

Pulse triggers transistor

Transistor would have triggered here

Fig. 7.31. Synchronization.

and

$$0 \cdot 7 R_4 C_2 = 0 \cdot 7 \times 40 \times 10^3 \times 220 \times 10^{-12} \, \text{s}$$

$$= 6 \cdot 16 \, \mu\text{s}$$

Mark-to-space ratio $= 2 : 1$ (or $1 : 2$)

Periodic time $= 3 \cdot 08 + 6 \cdot 16 = 9 \cdot 24 \, \mu\text{s}$

$$f = \frac{10^6}{9 \cdot 24} = 108\,200 \, \text{Hz or } 108 \cdot 2 \, \text{kHz}$$

Synchronization

The astable multivibrator is obviously of use as a generator of pulses or square waves. Its frequency is dependant on many factors, as was the case with sinusoidal oscillators, and a multivibrator is frequently synchronized with a train of pulses of an accurately known and stable frequency. This pulse train might itself be produced by some form of crystal oscillator. The frequency of the unsynchronized multivibrator is designed to be a little low, i.e. the periodic time is longer than that of the desired waveform. The synch pulses are fed onto one of the bases, and one such pulse is shown superimposed on the base waveform in Fig. 7.31. As can be seen the pulse forces the transistor into conduction earlier than it would otherwise have done. The pulse must, of course, be of sufficient height to push the base voltage up to about 0·6 V.

This procedure can be used to run an astable multivibrator at a frequency which is a submultiple of the p.r.f. Hence, in Fig. 7.32 the transistor starts to conduct on the fourth pulse and the multivibrator will run at exactly one quarter of the frequency of the pulses. Obviously the pulses, as well as being large enough for the fourth pulse to push the transistor into conduction, must not be so large that the third pulse does. This, so called, frequency division can be used without too much difficulty for divisions up to about ten times. Above this figure the pulse height becomes so critical that there is too much danger of triggering by the wrong pulse.

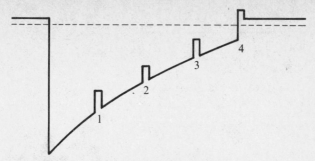

Fig. 7.32. Synchronization on fourth pulse.

Monostable multivibrator

A typical monostable multivibrator is shown in Fig. 7.33. T_1 is coupled to T_2 base as in the astable multivibrator, but the other coupling is different. When the circuit switches to the state with T_1 cut off and T_2 conducting it is, in fact, in a stable condition. T_2 collector will be low (near zero) and the potential divider formed by R_4 and R_5 will hold T_1 base negative (or cut off). If a positive input pulse raises T_1 base to about $0 \cdot 6\,\text{V}$, T_1 starts to conduct, T_1 collector falls and T_2 base falls leading to T_2 being cut off and T_1 conducting. However this state is not stable and, as in the astable multivibrator, C charges via R_3 till T_2 again conducts. Figure 7.34 shows the waveforms involved.

The effect is that an input pulse produces an output of length given by

(7.12) $$T = 0 \cdot 7 R_3 C.$$

The falling part of the output pulse can be used to trigger another pulse generator circuit, thus producing a pulse delayed by a time T with respect to the input pulse.

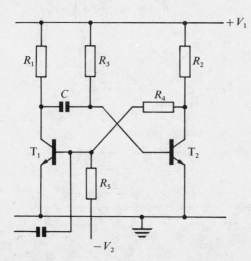

Fig. 7.33. Monostable multivibrator.

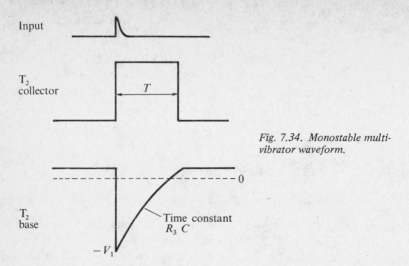

Fig. 7.34. Monostable multivibrator waveform.

Another common use of the circuit is in regeneration. Pulses being used in circuits like computers and telecommunication systems become somewhat distorted. The passage of pulses, for example, down a telephone line will distort them, and a monostable multivibrator can beused to generate new, clean, sharp pulses from the old, distorted ones.

Bistable multivibrator

The bistable uses two of the resistance couplings seen in the monostable, and an example is shown in Fig. 7.35. Either state is now stable. If T_1 is

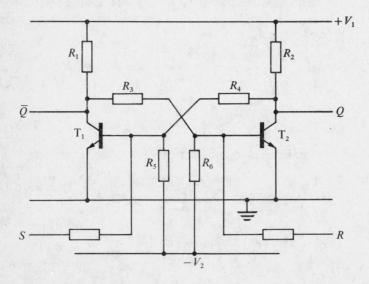

Fig. 7.35. Bistable multivibrator.

conducting, the fact that its collector is at about zero volts makes the base of T_2 negative (by the potential divider $R_3 R_6$) and holds T_2 off. Similarly with T_2 off the potential divider from $+V_1$ to $-V_2$ (R_2, R_4, R_6) is designed to keep the base of T_1 at about 0·6 V ensuring that T_1 conducts. Thus T_1 holds T_2 off and T_2 holds T_1 on.

A positive voltage applied momentarily to R will now cause T_2 to conduct and as its collector falls to zero it cuts T_1 off and the bistable has switched to its other state. Similarly a positive pulse applied to S switches it back to the original state.

This circuit is the heart of memory and counting circuits. It will be dealt with in more detail in the next chapter, where the meaning of the symbols S and R and where the reason for labelling the outputs (collectors) Q and $\overline{Q}$ will be explained.

7.C Test questions

1. Name, and show the waveforms produced by, the three basic multivibrator circuits.

2. An astable multivibrator, as shown in Fig. 7.29, has $R_3 = R_4 = 100\,\text{k}\Omega$. It is required to produce pulses of p.r.f. 500 kHz with a mark-to-space ratio of 1 to 5. Calculate the values of C_1 and C_2.

3. An astable multivibrator producing a square-wave output is to be synchronized at a p.r.f. of 100 kHz. If (in Fig. 7.29) $C_1 = C_2 = 50\,\text{pF}$ and resistors (for R_3 and R_4) and available with values of $130\,\text{k}\Omega$ and $150\,\text{k}\Omega$, which value would be chosen?

4. Explain why the input pulse height becomes critical in designing a synchronized multivibrator to divide the input frequency by more than a few to one.

5. Draw a circuit diagram of a monostable multivibrator. What is the length of the output pulse if the timing components are $100\,\text{k}\Omega$ and $50\,\text{pF}$?

6. Explain what is meant by pulse regeneration.

7. Draw a circuit diagram of a transistor bistable multivibrator. Explain why it is stable in either state.

8 Logic

8.1 Introduction

In previous chapters we have seen how information, such as speech or music, can be amplified by electronic devices. The signals being dealt with were very complex, and contained a great deal of information. Think of the wealth of detail contained in a few seconds transmission of a colour television picture! In this chapter we shall be dealing with a very different way in which information can be conveyed, and with a very different way of using semiconductor devices. We shall be considering how information can be contained in signals which have only two states.

Many devices can be thought of as having two states. A door may be open or shut; a switch may be on or off; a lever may be up or down; a diode or transistor may be passing a current or not. In the last example, we are not concerned with how much current is flowing but only with whether or not there is a current. Thus we are only concerned with the detection of a current, not with the measurement of its absolute value. In the circuits with which we shall be dealing the two states will be the presence of absence of a voltage.

It is possible to use two different voltages as the two states, but most logic systems use zero volts as one of the two states. As an illustration let us consider that the two states are 0 V and 6·0 V. The only two signals in our system are these two voltages. Slight differences will be ignored, so that if, for example, a signal is 5·7 V it will be considered as being 6 V. It is this very lack of the necessity to measure the voltages precisely that, perversely, makes this system so accurate.

It is possible to obtain any information by asking a series of yes/no questions. Thus, for example, a person's age can be obtained with any required degree of accuracy be questions such as 'are you over 40?', gradually narrowing down the time scale so that the age, even to the nearest second, could be found with sufficient questions.

All the information which the person has given is in two-state or binary form. In the field of communications it is possible to transmit very complex information, such as music or moving pictures, in the form of binary signals. Transmission of such information is beyond the scope of this book, but we shall deal with devices which are used to make logical decisions, and with circuits which are used for storing information.

8.2 The 'OR' gate

Consider, as a simple example, a chemical process in a factory, which is being monitored. The quantities being checked might be the temperature, the pressure, and the level of a liquid in a tank. It might be required to shut down the plant under certain conditions, for example if the temperature rises above 55°C, or if the pressure rises above 200 kPa, or if the liquid level falls below 2 m. Although these quantities could be measured by analogue devices we shall assume that they are measured by yes/no devices. The temperature might be monitored by a bimetallic strip which closes a switch at 55°C, the pressure by a pressure switch closing at 200 kPa, and the liquid level by a float which closes a switch when the level is below 2 m.

Each switch will be arranged so that in its closed position it switches 6 V onto a lead, the three leads going to a decision-making unit which will shut down the plant if this is needed. Although the decision-making unit could be a human operator, we shall, later in this chapter, be considering electronic circuits used for this purpose.

The type of problem just outlined is often put into mathematical notation. This is done not only because the problem can then be seen in a more concise form but also because mathematical techniques have been developed to enable simplifications easily to be made, if they are possible. We can represent the temperature rising above 55°C by the symbol 'A', a pressure rise above 200 kPa by the symbol 'B', and a liquid level of below 2 m by 'C'. We can now say that the plant must shut down if we have A or B or C. We can go further and represent a signal which will shut down the plant as, say, Z and thus we can say that we must have Z if we have A or B or C.

The simplification techniques which are discussed below form part of a branch of mathematics known as *Boolean algebra*, having been devised by George Boole (1815–64). The words 'if we have' in the statement above can be replaced by an equals sign and the word 'or' by the symbol '+'. (These were not, in fact, the symbols used by Boole, but are the ones now in use, and there is good reason for their choice.) Thus we may write the above statement as

(8.1) $$Z = A + B + C$$

We must be careful to read the right hand side of eqn (8.1) as 'A or B or C' and not as 'A plus B plus C'.

To perform the above logic function we require a circuit which will give, say, 6 V on its output (Z) if it has 6 V on one or more of its inputs (it being assumed, in this case, that the plant is to shut down not only if A or B or C is present but also if two or three of these conditions prevail). That is, we must have a 6 V output if there is 6 V on at least one input. Such a unit is known as an 'OR' gate.

The circuit symbols for an OR gate is shown in Fig. 8.1. Fig. 8.2 shows some superseded symbols which may still be encountered in use. The '1' in these figures is used because it is the minimum number of inputs which must be present to produce an output.

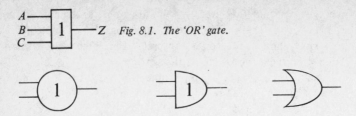

Fig. 8.2. *Some superseded 'OR' symbols.*

Not all logic systems use voltages of zero and 6 V as the two states, and it is more usual to use the symbols '0' and '1' to represent them. Thus an OR gate will give a '1' on its output for one or more '1's on its inputs. There are, in fact, eight possible input combinations with three inputs. In general, as each input can have two possible states, for n inputs there will be 2^n combinations. Figure 8.3 shows the eight input states for a gate with three inputs, together with the output, Z. This table, called by Boole a *truth table*, clearly shows the function of an OR gate.

8.3 The 'AND' gate

In the chemical process described above the plant was to be shut down for any one, or more, of the conditions listed. It might be that in a different situation we require it to shut down if the temperature rises above 55°C *and* if the pressure rises above 200 kPa *and* the liquid level falls below 2 m. In terms of the symbols we have used we require Z if we have A and B and C. In modern symbols the logic function 'AND' is shown as a dot, or full stop, hence

(8.2) $$Z = A.B.C.$$

In practice the dot is often omitted.

Figures 8.4, 8.5, and 8.6 show respectively the modern symbol for an AND gate, some superseded symbols, and the truth table.

A	B	C	Z
0	0	0	0
0	0	1	1
0	1	0	1
0	1	1	1
1	0	0	1
1	0	1	1
1	1	0	1
1	1	1	1

Fig. 8.3. *'OR' truth table.*

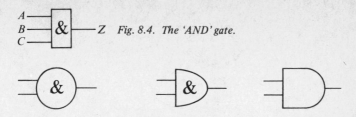

Fig. 8.4. The 'AND' gate.

Fig. 8.5. Some superseded 'AND' symbols.

8.4 The 'NOT' gate

In some situations we want a signal to exist (i.e. a '1' to be present) when we have NOT got some input condition present. For example, in our chemical process, we might require a green light to be on if the temperature is not above 55°C. Although this could be done by reversing the connections to the bimetallic strip, thus giving 6V or a '1' when the temperature is 55°C or below, we might already have made us of a '1' for a temperature above 55°C to shut down the plant in one of other of the previously outlined situations. In this case we simply use a gate which gives a '1' out for a '0' in and vice versa. It is called a NOT gate and the modern symbol is shown in Fig. 8.7 with some superseded symbols in Fig. 8.8. The truth table is given in Fig. 8.9. Note the circle on the output lead in Fig. 8.7. This is used in many logic diagrams to indicate negation. The modern symbol used in Boolean algebra is to put a line above the variable being negated, thus

(8.3) $$X = \overline{A}$$

8.5 Combinational logic

The examples which we have given of straightforward OR, AND, and NOT gates are relatively simple. Most practical problems require combi-

A	B	C	Z
0	0	0	0
0	0	1	0
0	1	0	0
0	1	1	0
1	0	0	0
1	0	1	0
1	1	0	0
1	1	1	1

Fig. 8.6. 'AND' truth table.

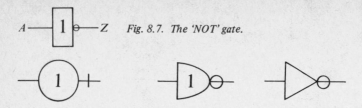

Fig. 8.7. The 'NOT' gate.

Fig. 8.8. Some superseded 'NOT' symbols.

nations of logic gates. For example, our plant may need to be shut down if we have a liquid level below 2 m and either a temperature above 55°C or a pressure above 200 kPa (or both). Thus

(8.4) $$Z = C.(A + B)$$

It is not very difficult to see that eqn (8.4) may be implemented using an AND gate and an OR gate as shown in Fig. 8.10. The following exercises will give you an introduction to combinational logic. Try them before you read the solutions.

Examples

Example 1

Draw logic diagrams showing the implementation of the following functions:
(a) $Z = A.B + C$
(b) $Z = A.B + C.D$
(c) $Z = A.B + \overline{C}$
(d) $Z = A + A.B$

Solutions for (a)–(d) are shown in Fig.s 8.11–8.14 respectively.

In (d) a little thought will show that both gates are redundant and that the statement is equivalent to $Z = A$.

A	Z
0	1
1	0

Fig. 8.9. 'NOT' truth table.

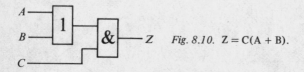

Fig. 8.10. Z = C(A + B).

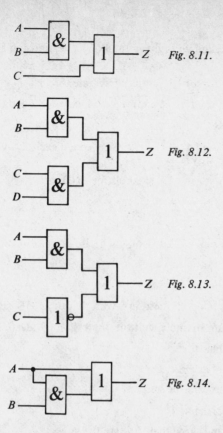

Fig. 8.11.

Fig. 8.12.

Fig. 8.13.

Fig. 8.14.

Example 2

A logic unit has two inputs, A and B. An output Z is required if the two inputs are different, i.e. if we have A and not B or if we have B and not A. Write down a Boolean expression for Z and devise a circuit to implement it.

Solution

$$Z = A.\overline{B} + \overline{A}.B$$

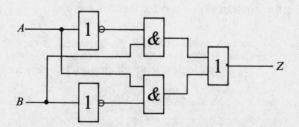

Fig. 8.15.

8.A Test questions

1. Draw a symbol for, and write down the Boolean expression for a three-input 'OR' gate.

2. Draw a truth table for a FOUR input 'OR' gate.

3. Construct a truth table for the expression of example 2 above, i.e.

$$Z = A.\bar{B} + \bar{A}.B$$

4. The logical expression $(A.\bar{B} + \bar{A}.B)$ is called an 'exclusive OR'. Can you give the reason why?

5. Sketch some of the superseded symbols for AND gates.

6. Draw a truth table for

$$Z = A + A.B.$$

7. Compare the truth table of question 6 with the statement $Z = A$.

8. Draw a truth table for

$$Z = A + \bar{A}.B.$$

9. Compare the truth table of question 8 with that of an OR gate.

10. Draw a logic diagram to implement the expression of question 8.

11. Construct truth tables for

$$A.(B + C) \text{ and } A.B + A.C$$

and hence show that they are the same.

12. Construct truth tables for

$$\overline{A + B} \text{ and } \bar{A}.\bar{B} \text{ and show that they are the same.}$$

13. Draw a truth table for $\overline{A.B}$ and show that it is the same as $\bar{A} + \bar{B}$.

14. A person is entitled to a certain type of grant if they are
    (a) a female aged 60 or over
or   (b) a married male.
To test the entitlement to a grant, three switches are provided:

    Switch A puts a '1' on its output if the person is 60 or over;
    Switch B puts a '1' on its output if the person is male;
    Switch C puts a '1' on its output if the person is married.

A light Z lights if the applicant is entitled to the grant.
Write down the Boolean expression for Z and devise a logical implementation.

15. A logic unit has two inputs, A and B. An output, Z, is required when either A and B are both '1's or both '0's. Write down the Boolean expression for Z and draw a logic circuit which would implement it.

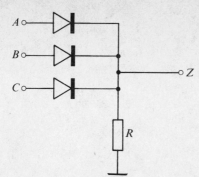

Fig. 8.16. Diode 'OR' gate.

8.6 Diode logic

We have, up to now, considered logic as a mathematical exercise. The physical realization of the three logic elements, or gates, described can be carried out in many ways, not necessarily electronic. The simplest electronic method is by the use of diodes. As an illustration we will consider that the two voltages are $+6$ V representing logic '1' and 0 V representing logic '0'. We will also assume that a non-conducting diode passes no current and that about 0·6 V is needed on its anode to cause it to conduct.

Fig. 8.16 shows a three-input diode 'OR' gate. If all three inputs are at logic '0' (i.e. connected to 0 V) the diodes are all nonconducting and the output, Z, will also be at 0 V or logic '0'. The resistor ties the output to earth (0 V) and its value is not very critical. If one or more of the inputs is connected to logic '1' ($+6$ V) the diode or diodes concerned will conduct and the output, Z, will be a little below $+6$ V. This slight drop in voltage is not significant in a simple circuit and the output will still be considered as logic '1'. The larger part of the 6-V input is dropped across the resistor R.

Fig. 8.17 shows a three-input diode 'AND' gate. Any one or more inputs at 0 V will cause the diode or diodes concerned to be forward biased, conducting and holding Z at almost 0 V (logic '0'). It is not until all

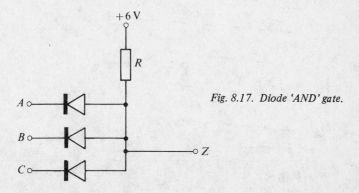

Fig. 8.17. Diode 'AND' gate.

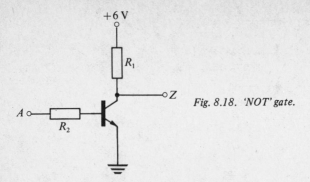

Fig. 8.18. 'NOT' gate.

the inputs are at $+6$ V that, because there is now no current in the resistor, Z becomes $+6$ V. Thus there is only logic '1' out when logic '1' is present on all the inputs, the property of an AND gate.

Although this section is headed 'diode logic' a transistor is needed to implement the NOT function, as shown in Fig. 8.18. If the input, A, is 0 V the transistor will be cut-off and the output, Z, will be at almost $+6$ V. Conversely when A is $+6$ V the transistor is hard-on, or saturated, and its collector voltage, the output, will be at about $0\cdot1$ or $0\cdot2$ V. Saturation is ensured by choosing a suitable value for R_2.

The action of the transistor in implementing the NOT function is really that of a switch. We are no longer trying to bias it in the centre of the load line, as was the case in amplifier design. Fig. 8.19 shows the output characteristics with a load line corresponding to R_1. When the transistor is 'On' the operating point, as can be seen, corresponds to a collector voltage of about $0\cdot1$ or $0\cdot2$, rather than the zero volts which would be dropped across a perfect switch. Similarly when 'Off' a small current is flowing, unlike a perfect switch which would pass no current when open.

Diode logic has the disadvantage that voltage levels change by the $0\cdot6$ or $0\cdot7$ V dropped across a conducting junction. This, as has been said, is not too important in simple circuits, but in a complex logic system the

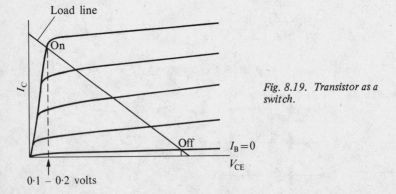

Fig. 8.19. Transistor as a switch.

cumulative effect may be to put the final output outside the permissible range of levels.

Before we look at ways of overcoming this difficulty, two terms, *'fan-out'* and *'fan-in'*, will be defined.

In explaining the action of the NOT circuit, Fig. 8.18, it was, in effect, assumed that there was no load on the output, Z. In fact, the output will normally feed other logic circuits with finite input resistances. Let us assume, as an example, that the input resistance of the other circuits being used is $20R_1$. If a NOT gate feeds such a logic unit, the output, Z, when indicating a logic '1' will, by potential-divider action, be $20/21 \times 6 = 5 \cdot 7$ V. If two such units are fed from the NOT gate giving a combined input resistance of $10R_1$, (two $20R_1$s in parallel), a '1' would be $10/11 \times 6 = 5 \cdot 45$ V, and so on. There is obviously a limit below which we can no longer say with certainty that the output is a '1'.

If this limit were, say, 5 V then, as 5 V would be produced if Z were loaded with $5R_1$, the number of inputs which it could feed would be limited to 4, giving an input resistance of $5R_1$. The circuit is said to have a *fan-out* of four. Similarly calculations could be carried out for other gates. Similarly the *fan-in* is the maximum number of inputs that a particular logic unit can have.

8.7 'NAND' and 'NOR' logic

The NAND gate

One way of overcoming the cumulative effect of the voltage drops in diode logic is to isolate gates from each other using transistor circuits of the type shown in Fig. 8.18. As this circuit produces the function NOT, an AND or OR gate would need to be followed by *two* such NOT gates to get back to the function AND or OR respectively.

However there are advantages, apart from the saving of a transistor and its associated circuitry, in using just one NOT gate. This is what is done in modern logic-circuits, whether made from discrete components or integrated circuits.

Consider an AND gate followed by a NOT as shown in Fig. 8.20. The signal at X will be $A \cdot B$ so that Z is given by

(8.5) $$Z = \overline{X} = \overline{A \cdot B}$$

which could be said to be NOT $(A$ and $B)$. In truth-table form X and Z are

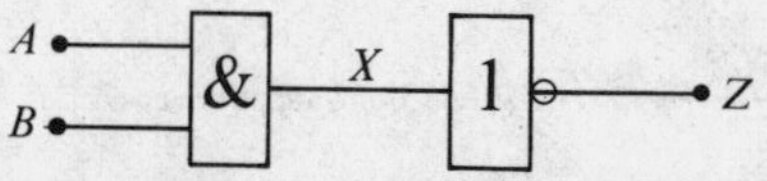

Fig. 8.20. NOT AND.

A	B	X	Z
0	0	0	1
0	1	0	1
1	0	0	1
1	1	1	0

Fig. 8.21. Truth table of Fig. 8.20.

seen in Fig. 8.21. This shows that Z will be a '1' unless A *and* B are both '1's, when it will be a '0'. Similarly for more than two inputs Z will only be '0' if *all* the inputs are '1'.

This function NOT AND is called NAND and needs rather more thinking about than the basic functions. Words like AND, OR, and NOT are used in everyday speech. NAND is not!

The Boolean expression for the NAND gate was shown in eqn (8.5) and the circuit symbol is seen in Fig. 8.22. The symbol is like the AND gate, but followed by the circle indicating NOT.

One advantage in using NAND gates is that all three basic logic functions may be implemented with NAND gates. In fact, we no longer need three types of gate, the NAND being sufficient.

A NOT gate can be made by using a NAND gate with one input. This gives a '1' out unless all the inputs (one in this case) are at logic '1'. As the NAND gates which you purchase would have a number of inputs, the question of what to do with the unused inputs arises. A look at Fig. 8.21 will soon make it obvious that unused inputs should go to logic level '1'. Most logic systems are such that an unused input assumes the value '1' if left on open circuit. However, it is normally better to connect unused inputs to the voltage level representing '1', usually via a resistor of a few kilohms.

Having made a NOT, an AND can be made, as in Fig. 8.23, using two NANDs. X is $\overline{A \cdot B}$ so that Z is $\overline{\overline{A \cdot B}}$ which is the same as $A \cdot B$.

To make an OR needs slightly more thought, the method being shown in Fig. 8.24. Each input is inverted and fed into a NAND. Obviously Z is $\overline{\overline{A} \cdot \overline{B}}$. Fig. 8.25 shows a truth table for this. Column Z will contain a '0' when, and only when, columns A and B *both* contain '1's. If we now compare Z with columns A and B we see a '1' in Z when there is a '1' in A or B (or both). This is the OR function.

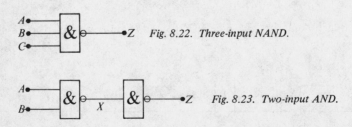

Fig. 8.22. Three-input NAND.

Fig. 8.23. Two-input AND.

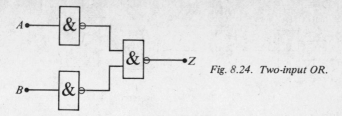

Fig. 8.24. Two-input OR.

A	B	$\overline{A}$	$\overline{B}$	$Z=\overline{\overline{A}.\overline{B}}$
0	0	1	1	0
0	1	1	0	1
1	0	0	1	1
1	1	0	0	1

Fig. 8.25. Truth table of Fig. 8.24.

This could have been seen by using the de Morgan theorem which you saw in the solution to question 13 on p.176. Applying de Morgan's theorem to $\overline{A}$ and $\overline{B}$, rather than A and B, gives

$$\overline{\overline{A}.\overline{B}} = \overline{\overline{A}} + \overline{\overline{B}} = A + B.$$

Thus any logic system of the type which we were considering earlier in this chapter can be made by replacing the ANDs, ORs, and NOTs by their NAND equivalents. (There are more sophisticated ways of designing NAND circuits which cannot be dealt with in this volume.)

The circuit of Fig. 8.10 which implements the equation $Z = C.(A + B)$ can be redrawn using NAND gates as in Fig. 8.26. Try to implement the functions of Example 1 (see page 174) before you look at the solutions.

(a) $Z = A.B + C$ (Fig. 8.11). (Solution shown in Fig. 8.27.)

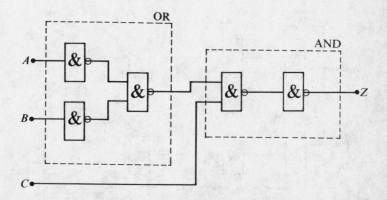

Fig. 8.26. Z = C (A + B).

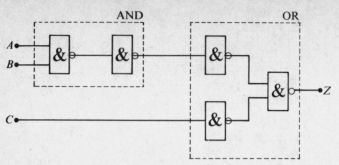

Fig. 8.27.

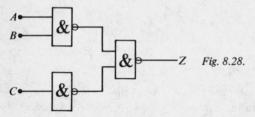

-Z   *Fig. 8.28.*

Now in this diagram there are two NOTs following each other; the second NAND of the AND feeds a NOT in the top input to the OR. These NOTs cancel, leading to Fig. 8.28.

(b) $Z = A.B + C.D$ (Fig. 8.12). (Solution shown in Fig. 8.29.) Leading to the diagram shown in Fig. 8.30.

(c) $Z = A.B + \overline{C}$. (Solution shown in Fig. 8.31.) Giving the diagram shown in Fig. 8.32.

(d) $Z = A + A.B$.
It has already been seen that this is equivalent to $Z = A$.

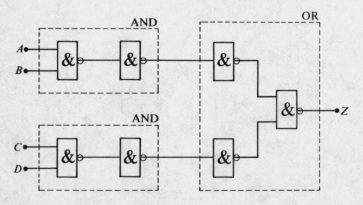

Fig. 8.29.

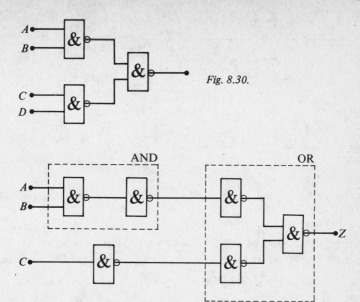

Fig. 8.30.

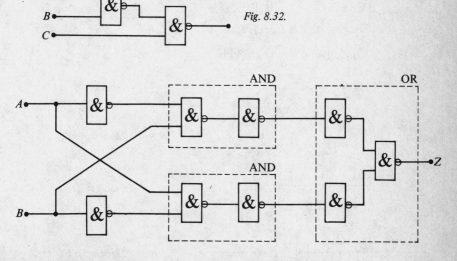

Fig. 8.31.

Example 2 was the exclusive OR or

$$Z = A.\overline{B} + \overline{A}.B.$$

Fig. 8.15 may be drawn in NAND form as shown in Fig. 8.33 which simplifies to the diagram shown in Fig. 8.34.

Fig. 8.32.

Fig. 8.33.

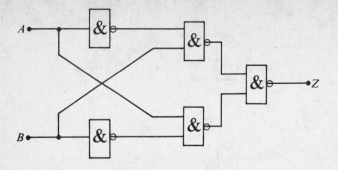

Fig. 8.34.

The NOR gate

In a similar way an OR followed by a NOT gives the function 'NOT OR' or NOR. The truth table is shown in Fig. 8.35 and the Boolean expression is

(8.6) $$Z = \overline{A + B}$$

Here the output is '0' unless *all* the inputs are '0', when it becomes '1'. The circuit symbol is given in Fig. 8.36.

As with the NAND gate, all three basic logic functions can be implemented from NOR gates and Figs. 8.37, 8.38, and 8.39 show, respectively, the functions NOT, OR, and AND implemented using NOR gates.

The circuit of Fig. 8.10 giving $Z = C(A + B)$ can be made from NOR gates as shown in Fig. 8.40. Fig. 8.41 shows a simplification of Fig. 8.40.

The three examples dealt with above can all be produced using NORs.

(a) $Z = A . B + C$ (Fig. 8.42).

(b) $Z = A . B + C . D$ (Fig. 8.43).

(c) $Z = A . B + \overline{C}$ (Fig. 8.44).

A	B	A+B	Z
0	0	0	1
0	0	1	0
1	0	1	0
1	1	1	0

Fig. 8.35. *NOR truth table.*

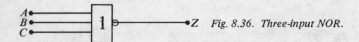

Fig. 8.36. Three-input NOR.

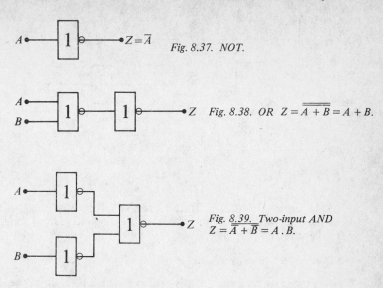

Fig. 8.37. NOT.

Fig. 8.38. OR $Z = \overline{\overline{A} + \overline{B}} = A + B.$

Fig. 8.39. Two-input AND
$Z = \overline{\overline{A} + \overline{B}} = A \cdot B.$

8.B Test questions

1. Draw circuit diagrams of a diode OR gate, a diode AND gate, and a transistor NOT gate. Explain their action.

2. A two input diode AND gate is shown in Fig. 8.45. If the two logic levels are 0 V representing '0' and +12 V representing '1', determine the output voltage (at Z) when A and B are both at logic '1' if Z is loaded with 20 kΩ.

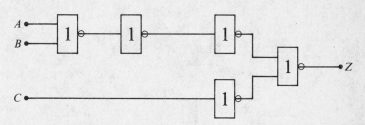

Fig. 8.40. $Z = C(A + B).$

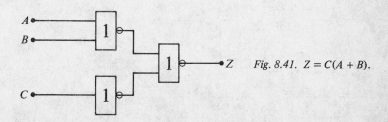

Fig. 8.41. $Z = C(A + B).$

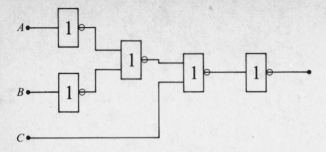

Fig. 8.42.

3. If, in question 2, logic '1' is defined as being +12 V ± 20 per cent, the number of units of input resistance 20 kΩ which can be connected to *Z* at the same time is

    (a) 4   (c) 1
    (b) 5   (d) any number

4. Describe how a transistor is used as a switch to perform the logic function NOT. Discuss, with the aid of a load line drawn on the output characteristics, in what ways the transistor departs from the perfect switch.

5. Draw a truth table showing the functions NAND and NOR for three inputs, *A*, *B*, and *C*.

6. Draw circuit diagrams showing how the three basic logic functions may be implemented using

    (a) NAND gates
    (b) NOR gates

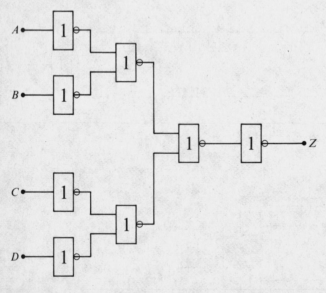

Fig. 8.43.

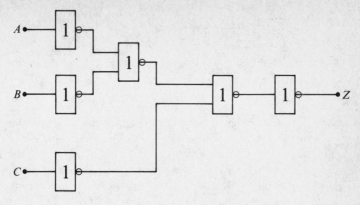

Fig. 8.44.

7. Implement the exclusive OR $(Z = A . \bar{B} + \bar{A} . B)$ using NOR gates.

8. Repeat question 14 on p.176 using
   (a) NAND gates
   (b) NOR gates

9. Repeat question15 on p.176 using
   (a) NAND gates
   (b) NOR gates

10. Fig. 8.46 shows a NAND logic circuit. Find Z in terms of A, B, and C. (Hint: you might find it easier to add in some NANDs to produce AND and OR circuits.)

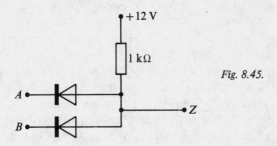

Fig. 8.45.

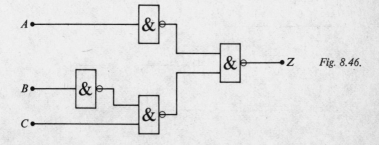

Fig. 8.46.

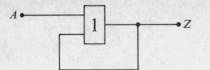

Fig. 8.47. An OR memory.

8.8 The R–S bistable

As soon as we progress past all but the most simple applications of logic, a *memory* is needed. When dealing with quantities which might alter with time (sequential as opposed to combinational logic) a circuit is required which *remembers* what signal was present at some previous time.

If a two input OR unit is made with one of the inputs connected to the output, Fig. 8.47, the circuit will remember if a '1' has ever been present on the other input. Assume that originally A is '0' and Z is '0'. If now A becomes '1', Z will become '1' (only one input is needed to produce an output) and even if A goes back to '0' the fact that Z is '1' keeps the other input at '1'. Hence $Z = $ '1' indicates that at some time A has been at logic '1'. *It remembers.* Z will now stay at logic '1' unless either the feedback path is broken, or the unit is switched off.

A more useful circuit can be obtained by replacing the OR unit by two NORs, Fig. 8.48. The input is now called S (for a reason that will be explained later) and a second input is taken to the other NOR. This input is called R. Further, it is conventional in this type of circuit to call the output Q.

Let us assume that, to start with, both S and R are at logic level '0'. Theoretically Q could be at '0' or '1' and the circuit is stable in either case. If Q is '0', then two '0's into NOR gate A produce an output, X, of '1' and a '0' and a '1' into NOR gate B gives a '0' out (Q). Similarly if Q is '1', X will be '0', keeping Q at '1'.

As the circuit is stable in either of two states it is called a *bistable*.

If the NOR gates are absolutely identical, there will be equal chances of Q becoming '0' or '1' when the unit is switched on. If Q is '0' a '1' applied momentarily to S will change Q to '1' as follows. NOR gate A has a '0' and a '1' on its inputs making X become '0' so that unit B has two '0's on its inputs ensuring that Q is a '1'. Q will stay at '1' even after the '1' is removed from S. Similarly if Q is '1' a '1' applied to R will force Q to become '0'.

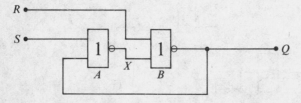

Fig. 8.48. R–S *bistable.*

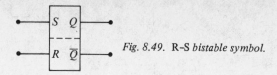

Fig. 8.49. R–S *bistable symbol.*

Hence the unit *remembers* whether a '1' was last applied to S or R. It should be noticed that X is *always* the inverse of Q and, for this reason, is called $\overline{Q}$. The circuit is rather like a see-saw which tips one way and then the other, and is often called a *flip–flop*.

By convention, if Q is '1' the circuit is said to be *set* and if Q is '0' it is said to be *reset*. Thus a '1' on S *sets* the bistable, a '1' on R *resets* it. Note that a '1' on both S and R is not an allowable condition (it leads to both Q and $\overline{Q}$ being '0'!) and logic circuits can easily be designed so that it is not possible to put a '1' on S and R at the same time.

Fig. 8.48 is not a very good diagram because it tends to give the impression that unit A is the input and unit B the output. In fact, the circuit is symmetrical and the modern symbol is shown in Fig. 8.49. A truth table for the operation of an R–S bistable is given in Fig. 8.50. It is, of course, a circuit which changes with time and hence the columns Q (the *previous* state) and Q' (the *new* state).

Rows 1 and 2 show that if $S = R = $ '0', Q does not alter. Rows 3 and 4 show that if $S = $ '0' and $R = $ '1', Q goes to, or stays at, '0'. Similarly if $S = $ '1' and $R = $ '0', Q goes to, or stays at, '1'. Lastly rows 7 and 8 show that $R = S = $ '1' is not an allowable state.

A similar circuit can, of course, be constructed using NANDs, Fig. 8.51. The difference is that the non-allowable state is now $S = R = 0$. If $S = R = 1$ the bistable remains in whatever state it is in, and, as the truth table of Fig. 8.52 shows, a 1 on S with a 0 on R *sets* the bistable, whereas a 0 on S and a 1 on R *resets* it. Note that these two conditions for setting and resetting the device are the same as for the bistable made from NOR gates.

	S	R	Q	Q'
1	0	0	0	0
2	0	0	1	1
3	0	1	0	0
4	0	1	1	0
5	1	0	0	1
6	1	0	1	1
7	1	1	0	X
8	1	1	1	X

Fig. 8.50. R–S *bistable truth table.*

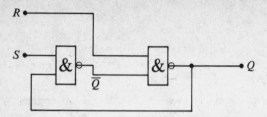

Fig. 8.51. R–S *bistable using NANDs.*

S	R	Q	Q'
0	0	0	X
0	0	1	X
0	1	0	0
0	1	1	0
1	0	0	1
1	0	1	1
1	1	0	0
1	1	1	1

Fig. 8.52. R–S *bistable (NANDs) truth table.*

A very simple use of the R–S bistable is as a memory in a push-button switch controlling, say, a large motor, Fig. 8.53.

Q will, if at logic '1', operate a device, such as a thyristor, which switches power to the motor. Pushing the 'ON' button sets the bistable, putting a '1' on Q and starting the motor. Similarly the 'Off' button resets the bistable, or puts a '0' on Q.

In practice we could incorporate some form of safety circuit, if required, in case an operator pushes both buttons at the same time. One possible circuit is given in Fig. 8.54. The NAND gate detects the condition of both buttons pressed, as shown. It gives a '0' in this condition, which stops the 'On' signal passing through the AND gate. This ensures that in this event the bistable will be RESET. (The AND would, of course, consist of two NANDs.)

The bistable was dealt with in the previous chapter and Fig. 7.35 (p.168) shows a circuit diagram. In effect each transistor is a two input NOR gate, one input coming from S or R and the other from the other transistor.

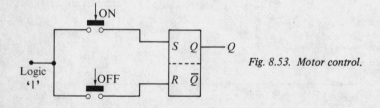

Fig. 8.53. Motor control.

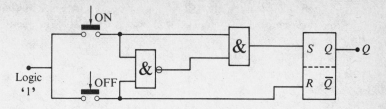

Fig. 8.54. Motor control with safety interlock.

8.C Test questions

1. Discuss the differences between combinational and sequential logic.

2. An R-S bistable is to be made using two NANDs. It is essential to SET and RESET it using logic '1' on S and R respectively. Draw the circuit (block) diagram and explain its action. Also draw the circuit symbol of an R-S bistable.

3. The basis of counting circuits is the bistable. It is arranged such that if a pulse (a '1') is applied to an input, it is steered to R if Q is at '1' and to S if Q is at '0'. Hence whichever state the bistable is in, the pulse will cause it to change to the other state. Design a possible circuit using a bistable and two AND gates. Remember that the $\overline{Q}$ output is available for use as well as the Q output.

8.9 Digital integrated circuits

A brief outline of linear ICs was given in Section 6.6. The advantages of ICs given there apply equally to digital circuits.

Details of the circuitry of logic gates, other than very simple diode logic, have not been given in this book. However, two main types of gate are available in IC form. In TTL circuitry (that is circuitry based on bipolar transistors), chips are available with NAND gates having various numbers of inputs. One chip may have 4 two input gates, 3 three input gates, 2 four input gates, or 1 eight input gate. The present price (1978) is between 20 and 30p for a single chip. Other price examples are:

monostable multivibrator	55p
decade counter	£1.90
timer (from ms to days)	£3.90

Digital ICs are also available using MOS transistors. These have the advantages of smaller size, easier fabrication, and high input resistance. Against these advantages are the slower speed of operation, the lower voltage breakdown, and the higher noise level. At present chips containing 4 two-input gates cost about 20p each. A complete counter costs about £1 and a J-K bistable (much more complex and versatile than the R-S bistable) about 50p.

It should be noted that all the prices quoted here and in Chapter 6 are for single chips. Large reductions in price are available for quantity orders.

9   Valves

9.1 Introduction

The thermionic diode was first used by Fleming at the beginning of the century. It was used as a device which would pass current between two electrodes in one direction but not in the other—hence the name valve. Soon after, De Forest discovered that a third electrode, the grid, enabled the device to amplify, and electronics had arrived. The tetrode, pentode, and many other valves evolved. It would indeed be an understatement to say that the art of communication was revolutionized.

As often happens, there was enormous technical advance due to war, in this case the Second World War. Modern miniature valves were developed, a type of valve which is still in use in some electronic equipment, including television sets.

Just how important the valve is to your particular course of study is hard to say. Many technicians are still involved in the maintainance of valve equipment, and could well be for some years to come. There are some areas where valve equipment is still being designed. One example is in large transmitters, although in this case the valves are not quite like those that we shall be considering, at least in mechanical construction.

9.2 Electron emission

In §2.1 the concept of free electrons moving about in a substance at random was discussed. Under certain circumstances it is possible for some of these electrons to escape from the surface of the material. However, to do so they must be given energy. There is a barrier to be overcome, because as an electron leaves the surface of the material the positive charge left behind, due to the absence of the electron, will try to pull it back. It is rather as though the electron is a ball and the barrier a hill up which it must roll to escape.

The energy required varies from substance to substance. It is usually measured in electron-volts (eV), the electron-volt being the kinetic energy which an electron would gain in falling through a potential of one volt. As the charge on the electron is $1·6 \times 10^{-10}$ C an electron-volt must be $1·60 \times 10^{-19}$ J. The amount of energy needed for escape is a very important parameter of a substance, and is called its *work function*.

There are various ways in which an electron can acquire sufficient energy to escape. We are concerned here mainly with the thermal energy given if the substance is heated, but it is worth mentioning other ways. Certain substances emit electrons if radiation of particular wavelengths falls on them. This type of emission is called *photoelectric emission* and is used in photoelectric cells. Energy may also be given by the actual bombardment of the material with a stream of fast moving electrons or other particles. The energy of the bombarding particles is transferred to electrons within the material. Some of these electrons acquire sufficient energy to overcome the potential barrier at the surface. Such electrons are called secondary emitted electrons. Their effect is often undesirable as in the tetrode, section 9.6. However, as is often the case in engineering, what is undesirable in one context can be made use of in another, and secondary emission is made use of in photomultipliers. Lastly electrons can be extracted by applying a strong enough electric field which, so to speak, tears them out of the material. This is called *field emission*.

Thermionic emission

Most metals when heated melt before a temperature is reached at which they emit electrons. One in which this does not happen in tungsten, with a work function of 4·54 eV. If tungsten is heated to about 2500 K, well below its melting point of 3655 K, it can produce an electron current of 650 mA/cm^2 of surface. Perhaps a more important parameter of a cathode is its efficiency, defined as emission current in amperes per watt of heating power. For tungsten it is between 10 and 100 mA/W. This, in fact, is not particularly good and the use of pure tungsten tends to be limited to high-voltage valves.

The reason why it is necessary to use pure tungsten for the filaments of high-voltage valves is both interesting and important. No vacuum is perfect and there will be, inside the valve, some atoms of various gases. The emitted electrons, which are moving towards the anode, will collide with gas atoms, perhaps knocking out one of the electrons circling the nucleus. If this happens the atom is left with a positive charge and is called a positive ion, the effect being called ionization. The positive ion will move towards the cathode and, in a high-voltage valve, the ions will be moving fast enough to damage it. (The acceleration of the ions in the valve will be proportional to the anode voltage.) Pure tungsten is the only emitting material in use which will withstand this bombardment.

It has been found that if 1 or 2 per cent of thorium oxide is added to the tungsten it can be heated to temperatures well above that at which pure thorium would evaporate. This *thoriated tungsten* has a work function of 2·63 eV and at 1900 K will give 1·5 A/cm^2. Its efficiency is 50–1000 mA/W—much higher than pure tungsten. It is easily damaged by positive-ion bombardment however, and its use is limited to valves operating with anode voltages below 5000 V.

Anode voltages of thousands of volts are only encountered in rather special applications and a large proportion of valves operate with anode voltages in the order of a few hundred volts. The cathodes used are *oxide coated*, consisting of a nickel tube coated with barium and/or strontium oxides. Unlike the tungsten and thoriated tungsten filaments, where the heating current is passed through the material, the oxide coated cathode is indirectly heated by a separate heater, insulated electrically from the cathode. This type of cathode will be discussed further in the next section. All we need point out here is that the work function of these materials is about 1 eV and that at 1000 K such a cathode can emit up to $10\,A/cm^2$, the efficiency being as high as 10 A/W. They are very delicate and their use restricted to low voltage applications.

9.3 The vacuum diode

The diode, as explained earlier, is a device which will pass current in one direction only. Electrons emitted by a cathode move towards a second electrode, the anode, under the influence of the electric field between the electrodes when, and only when, the anode is positive with respect to the cathode. The current is thus a flow of electrons from cathode to anode, although this is a flow of *conventional* current from anode to cathode.

As mentioned in the previous section, there are two types of cathode, directly heated and indirectly heated.

The directly heated cathode is nowadays almost entirely confined to high- or medium-voltage valves. The heating current is usually d.c., often provided by a 1·5-V battery or other supply. The cathode, or filament, is made of tungsten or thoriated tungsten, although some oxide-coated directly heated filaments are used (only, of course, for low-voltage applications).

The indirectly heated cathode (Fig. 9.1) consists of a nickel tube coated with the oxides, and heated by a tungsten filament insulated from the cathode with aluminium oxide. As the heater is not electrically connected to the cathode (or anode) it is possible to use a.c. to heat it. In fact most modern small valves use either 6·3 V a.c. supplied by a mains transformer or, in the case of circuits such as television receivers, they have heaters wired in series together with a suitable resistor and are connected directly

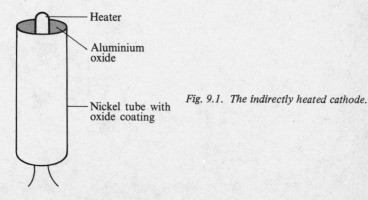

Heater

Aluminium oxide

Nickel tube with oxide coating

Fig. 9.1. The indirectly heated cathode.

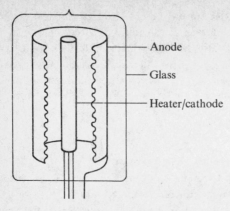

Anode

Glass

Heater/cathode

Fig. 9.2. Construction of a diode.

across the mains supply. The latter method, although having disadvantages, avoids the need for a mains transformer, which is quite an expensive item.

The other electrode, the anode, consists of a nickel tube, often cylindrical, surrounding the cathode (Fig. 9.2). The assembly is sealed into an evacuated container, usually glass, and the four connections, two heater leads, one cathode, and one anode lead, protrude through the bottom in the form of pins which enable the valve to be plugged into a socket, or valve holder. The circuit symbols for directly and indirectly heated diodes are shown in Fig. 9.3.

Fig. 9.4 shows a circuit suitable for determining the characteristic of a

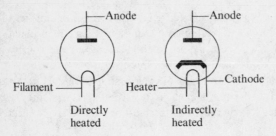

Anode   Anode

Filament   Heater   Cathode

Directly   Indirectly
heated   heated

Fig. 9.3. Symbols for diodes.

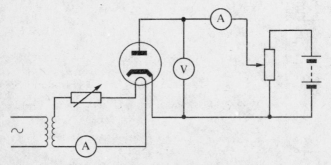

Fig. 9.4. A circuit for determining diode characteristics.

diode (i.e. a graph of anode current against anode voltage). The characteristic of a typical small diode is drawn in Fig. 9.5.

The first part of the curve, marked *OA*, is somewhere between a straight line and a square-law curve. In theory over this region $I_a \propto V_a^{\frac{3}{2}}$, this law being known as the three-halves-power law. At higher anode voltages the characteristic becomes almost horizontal or *saturated* as shown, because here the anode is collecting all the electrons which the cathode can emit at its particular temperature. The circuit shown includes a means of altering the heater current and hence the cathode temperature, and Fig. 9.5 shows a characteristic for a reduced heater current. However, in normal diode applications the heater current is kept constant.

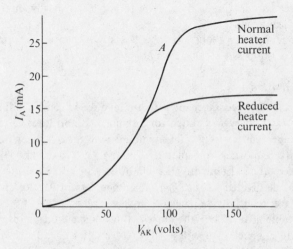

Fig. 9.5. *Diode characteristics.*

It might be thought at first that as long as the anode is positive it will attract all the electrons emitted by the cathode, in which case the characteristic would be a vertical line coincident with the current axis (becoming horizontal, of course, at the saturation current). This would be an ideal diode, dropping no voltage. The reason for the curved characteristic is that there exists what is known as a *space charge*.

As the cathode heats up electrons emitted from it form a cloud or space charge round it. This negative charge actually repels electrons back into the cathode, and, with no anode voltage applied, an equilibrium is established with just as many electrons returning to the cathode as are emitted from it. As the anode voltage is increased positively it attracts electrons from the space charge, the electrons being replaced by newly emitted ones, until, at a high enough voltage, it is attracting as many electrons as the cathode can emit.

The two distinct regions of the characteristic are often described by saying that the current is *space-charge limited* in the region *OA* and *temperature limited* for saturation.

Obviously the anode current will be zero if the anode voltage is negative and hence we have a valve.

The main differences between the thermionic and semiconductor valve are:

(a) the reverse current of the thermonic valve is zero unlike the semi-conductor which has a very small reverse current;
(b) the thermionic valve drops a larger voltage at a given current and hence dissipates more power;
(c) the semiconductor diode requires no heater supply;
(d) the semiconductor diode is more robust mechanically but far less so electrically; and
(e) a semiconductor diode occupies less space than a thermionic diode designed to pass the same current.

Diode resistance

The a.c. and d.c. resistance of a thermionic diode can be defined in the same way as they were defined for a semiconductor diode. The static or d.c. resistance will vary by a large amount with the operating point and is not a very important parameter. The dynamic, a.c., anode, or plate resistance (plate is frequently used in American books) is the *change* in anode voltage divided by the *change* in anode current. It is often relatively constant over a fairly wide range of anode voltages.

The problem of finding the current flowing if a diode, a resistor, and a battery are connected in series can be solved by a load line construction exactly as for a semiconductor diode. Such an exercise will be found in the test questions on this section.

Diode ratings

A number of points must be observed when designing a diode circuit. Firstly, the heater must be operated at the specified voltage or current. The diode, like the semiconductor, must operate without the stated maximum current, voltage, or power being exceeded. If it operates at too high a power the anode may rise to such a temperature that it emits gases destroying the vacuum or may even melt. Some anodes may, however, operate at a dull red colour without undue emission of gas. Whilst operation above the ratings must not be considered, an accidental temporary overload, even quite a large one, will probably not damage the thermionic diode whereas quite a small overload for even a fraction of a second might irreparably damage a semiconductor device.

Like the semiconductor diode there will also be a peak inverse voltage which must not be exceeded, although it will be considerably higher. In valves designed for very high voltages the peak inverse voltage might be as high as 200 000 V.

Lastly, for indirectly heated diodes, there is a maximum voltage which is allowed between heater and cathode. This will be considered in more detail for the triode.

9.A Test questions

1. The electron-volt is a measure of
    (a) the voltage across a valve
    (b) the current in a valve
    (c) energy
    (d) power

2. List four methods of causing certain metals to emit electrons.

3. List the three common cathode materials in order of efficiency.

4. Tungsten is used as a cathode in high-voltage valves because
    (a) it has a higher work function than the other materials in use
    (b) it is mechanically stronger than the others
    (c) it can withstand the bombardment by secondary electrons

5. The reverse current in a thermionic diode is
    (a) negligible or zero
    (b) about of the same order as in a semiconductor
    (c) slightly more than in a silicon diode

6. Draw a diagram of a circuit which could be used to find the anode characteristics (anode current/anode voltage) of a thermionic diode.

7. Sketch a typical anode characteristic of a thermionic diode, showing saturation. Draw, on the same sketch, a curve for the same diode operating at a higher cathode temperature and explain the difference between the two curves.

8. The space charge in a thermionic diode is
    (a) a cloud of electrons round the cathode
    (b) a cloud of positive ions round the cathode
    (c) produced because the vacuum is not perfect
    (d) only present in a gas-filled valve.

9. Temperature-limited operation is
    (a) that part of the characteristic where $I_a \propto V_a^{\frac{3}{2}}$
    (b) that part of the characteristic where $V_a \propto I_a^{\frac{3}{2}}$
    (c) saturation
    (d) that part of the characteristic where the voltage across the valve is almost independent of current.

10. List four differences between semiconductor and thermionic diodes.

11. The diode whose characteristic is shown in Fig. 9.5 is connected in series with a 7·5 kΩ resistor to a 150-V d.c. supply. Draw a load line and hence estimate
    (a) the current flowing
    (b) the voltage across the diode

(c) the voltage across the resistor

Assume that the normal heater current is flowing.

12. Which is the higher for a thermionic diode
(a) the static (or d.c.) resistance
(b) the dynamic (or a.c.) resistance.

13. Define the dynamic (a.c.) resistance of a diode.

14. Estimate both the d.c. and a.c. resistances of the diode whose characteristic is shown in Fig. 9.5 at the operating point of question 11.

15. List the various ratings of a thermionic diode which you would need to consider in choosing a valve for a particular application.

9.4 The triode

In 1906 De Forest, an American, patented the idea of introducing a third electrode into the diode. It is called the *grid* and consists of a widely spaced wire mesh surrounding, and positioned just outside the cathode (Fig. 9.6).

If the valve, called a *triode*, is operated with the grid connected to the cathode, the characteristic will be virtually the same as for a diode. However, if the grid is made negative with respect to the cathode, then the anode voltage has difficulty in attracting electrons from the space charge, which means that the anode current will be lower than if the grid were at cathode potential. Thus the grid voltage has a large influence on the current through the valve, just as the gate voltage of a FET controls the drain current. Note that in normal operation the triode is not operated with the grid positive with respect to the cathode as this would cause current to flow to the grid.

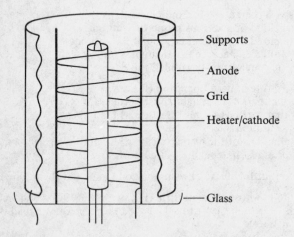

Fig. 9.6. Grid and cathode of a triode.

As an example of the effect of the grid in controlling the anode current consider that the characteristics of a triode are such that with the grid connected to the cathode (i.e. $V_{GK} = 0\,V$), an anode voltage of $50\,V$ produces an anode current of $10\,mA$. If the grid is now made one volt negative ($V_{GK} = -1\,V$) the anode current will fall. If the anode voltage is now increased the current will rise. Let an increase of $30\,V$ on the anode raise the current back to its original value, $10\,mA$. This increase in anode voltage which just counteracts the negative change of one volt on the grid is an important parameter of the triode, known as the *amplification factor* and is given the symbol μ (mu). (A similar parameter can be defined for the FET.) Hence we have

(9.1)
$$\text{Amplification factor } \mu = \frac{\text{Positive change in anode voltage}}{\text{Negative change in grid voltage}}$$

with anode current remaining constant.

With the above figures μ has a value of 30. The reason that it is greater than unity is that the grid, because it is much closer to it, has a greater effect on the current leaving the cathode than does the anode. Values of μ between about 5 and 100 are typical for a triode, the actual value being a function of the dimensions of the various electrodes.

Output characteristics

Just as for the transistor and the FET it is very important to know what is happening at the output of the valve, that is, how the anode current varies with anode voltage. The output characteristics, sometimes called the anode or plate characteristics, may be plotted from readings of anode current and voltage at a constant value of grid voltage. These readings may be taken with the circuit shown in Fig. 9.7 and a set of output characteristics for a small triode is given in Fig. 9.8. Fig. 9.7 also shows the circuit symbol for an indirectly heated triode, the dotted line representing the grid.

There is no need to measure grid current because, with the grid negative, this current will be negligible.

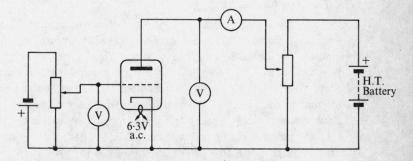

Fig. 9.7. Circuit for determining triode characteristics.

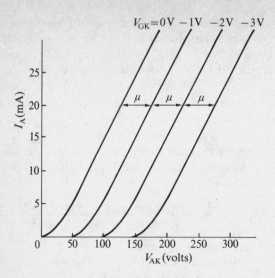

Fig. 9.8. *Triode output characteristics.*

It should be obvious that if the anode voltage is increased sufficiently saturation would occur. This has not been shown on the output characteristics because it is not usual to operate a triode in this region.

The curves shown in Fig. 9.8 are drawn for values of grid voltage 1 V apart. They must then be separated by μ anode volts as shown, this following from the definition of μ (eqn 9.1).

The reciprocal of the slope of the output characteristics represents the change in anode voltage required to produce unity change in anode current, and, this is defined as the dynamic output resistance. It is given the symbol r_a and hence

(9.2)
$$r_a = \frac{\text{Change in } V_{AK}}{\text{Change in } I_A} \text{ at a constant } V_{GK}.$$

Values of r_a for a triode vary from about $0.5\,\text{k}\Omega$ to $100\,\text{k}\Omega$.

Transfer characteristic

Again, as for a transistor or FET, a graph of output against input, in this case anode current against grid voltage, is called the *transfer characteristic*. Readings can be taken using the circuit of Fig. 9.7 or can be ascertained from the output characteristics, Fig. 9.8. However, the anode voltage has a great effect on the transfer characteristic and a set of such graphs for different anode voltages is shown in Fig. 9.9. They are, like the output characteristics, equally spaced and fairly straight over a wide range of V_{GK}.

It will be seen that for any particular anode voltage if the grid voltage is made sufficiently negative anode current ceases to flow. The valve is then said to be *cut-off*.

The slope of the transfer characteristics is called the mutual conductance, g_m, as for the FET and

(9.3) $$g_m = \frac{\text{Change in } I_A}{\text{Change in } V_{GK}} \text{ at a constant } V_{AK}.$$

For a triode g_m ranges from about 0·5 to 20 mA/V (or mS).

Relationship between parameters

r_a, g_m, and μ are the three parameters of the triode. From eqns 9.1, 9.2, and 9.3 it can be seen that

$$r_a \times g_m = \frac{\text{Change in } V_{AK}}{\text{Change in } I_A} \times \frac{\text{Change in } I_A}{\text{Change in } V_{GK}}$$

$$= \frac{\text{Change in } V_{AK}}{\text{Change in } V_{GK}} = \mu.$$

(9.4) Thus $r_a g_m = \mu$.

For example, if $r_a = 20 \text{k}\Omega$ and $g_m = 10 \text{mA/V}$ then $\mu = 20.10^3 \times 10.10^{-3} = 200$.

Triode ratings

Like the diode the triode will have a maximum anode current, anode-cathode voltage, and also a maximum power dissipation. It cannot be operated at its maximum anode current and voltage at the same time because the maximum power would then be exceeded.

Although the grid is normally operated at a negative voltage there are circumstances when a positive voltage may be applied and the ratings will

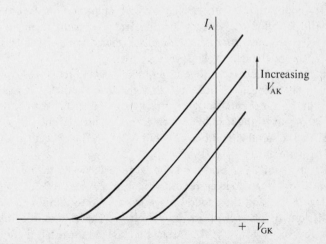

Fig. 9.9. Mutual or transfer characteristics of a triode.

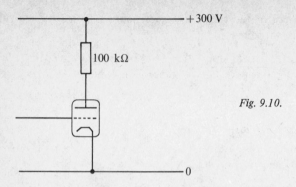

+300 V

100 kΩ

Fig. 9.10.

0

include a maximum value above which the current flowing to the grid would cause damage.

Like the diode, triodes with an indirectly heated cathode will have a specified maximum heater–cathode voltage. This is often relatively low, in the order of 100 to 150 V. If the insulation between the heater and cathode were too good it would impede the flow of heat. It is usually very desirable to earth one side of the heater or to connect it to the cathode to prevent pick-up of the 50-Hz heater supply in the valve itself, resulting in hum. When a transformer is used to supply the heaters, a number of heaters may be run from one secondary winding. Care must be taken because all the heaters will then be at the same d.c. potential and, if some of the cathodes are at different potentials to others, and hence at different potentials to the heaters, breakdown in heater–cathode insulation can result.

9.5 The triode amplifier

Valve amplifiers are very similar to FET amplifiers—in fact the FET is sometimes called the semiconductor valve! The main differences are that the valve operates from a considerably higher d.c. supply and also usually has a very much higher voltage gain.

Figure 9.10 shows a triode in series with a load resistor. Assuming that the supply voltage is 300 V and that the load is 100 kΩ, a load line has been drawn on the output characteristics of Fig. 9.11. With a triode there is a large part of the load line which cannot be used (from A to the current axis) and a suitable grid bias is about −2·0 V. The Q point is defined by $I_A = 1·5$ mA and $V_A = 150$ V.

An input of 2·0 V peak or 4 V peak to peak swings the operating point from A to B, the anode current varying from 2·4 mA to 0·75 mA (1·65 mA peak to peak) and the anode voltage from 57 V to 225 V (168 V peak to peak), out of phase with the input voltage. This results in a voltage gain given by $A_v = 168/4 = 42$. As in the case of the FET there is no real meaning to current gain.

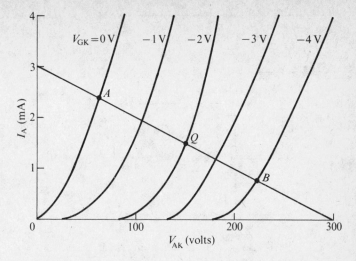

Fig. 9.11.

Bias may be provided by a cathode resistor which lifts the cathode, in this case, to $2{\cdot}0\,\mathrm{V}$ positive, the grid being earthed via a large value resistor as shown in Fig. 9.12. The value of the cathode resistor is $2{\cdot}0\,\mathrm{V}/1{\cdot}5\,\mathrm{mA} = 1{\cdot}3\,\mathrm{k\Omega}$.

The capacitor C, which keeps the bias constant with the varying anode current, should have a reactance of about 10 per cent of the value of the cathode resistor at the lowest frequency to be amplified. If the latter is $40\,\mathrm{Hz}$

$$C = \frac{1}{2\,.\,\pi\,.\,40\,.\,130}\,\mathrm{F} = 31\,\mu\mathrm{F}.$$

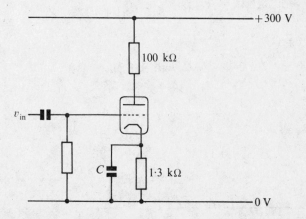

Fig. 9.12. Cathode bias.

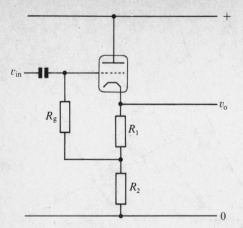

Fig. 9.13. A cathode follower.

The valve is also used in the common-grid and common-anode (cathode-follower) configurations. The latter is commonly used for matching a high resistance source to a low resistance load, such as a cable. Fig. 9.13 shows a typical cathode-follower circuit. $(R_1 + R_2)$ is the load, but because this would provide too high a bias, the grid is returned to the junction of R_1 and R_2, R_1 being calculated to provide the required bias.

9.6 The triode equivalent circuit

The equivalent circuit of a triode is really the same as that of an FET, except that we are not able to ignore the slope resistance, r_a, as we sometimes ignored r_d for the FET.

Using the characteristics of Fig. 9.11 we can estimate values of g_m and r_a (at Q) of about $1.7 \, \text{mA/V}$ and $33 \, \text{k}\Omega$ respectively. Hence we get the equivalent circuit of Fig. 9.14 for the circuit of Fig. 9.12. Now, $33 \, \text{k}\Omega$ in parallel with $100 \, \text{k}\Omega$ gives

$$\frac{33 \times 100}{33 + 100} = 24.8 \, \text{k}\Omega.$$

Hence $A_v = 1.7 \times 24.8 = 42.2$, with $180°$ phase shift.

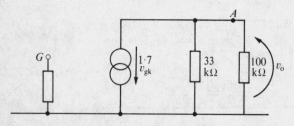

Fig. 9.14. Equivalent circuit of Fig. 9.12.

Lastly, mention should be made of another frequently used equivalent circuit of the valve which depicts the valve as a voltage generator in series with a resistance. This equivalent circuit of the triode amplifier is shown in Fig. 9.15. The polarity of the generator indicates the 180° phase shift.

From the equivalent circuit, the output voltage is given by

$$v_o = \frac{R_L \mu v_{gk}}{r_a + R_L}$$

and

(9.5)
$$A_v = \frac{v_o}{v_{gk}} = \frac{\mu R_L}{r_a + R_L}.$$

Now

$$r_a g_m = 33 \times 1\cdot7 = 56\cdot1.$$

Hence

$$A_v = \frac{56\cdot1 \times 100}{33 + 100} = 42\cdot2$$

as before.

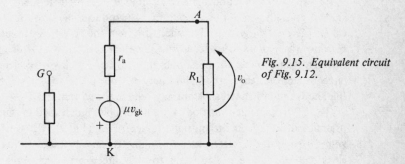

Fig. 9.15. Equivalent circuit of Fig. 9.12.

9.7 Other valves

It is not intended to devote much space to a consideration of other valves because the principles on which they operate are in effect the same as for a triode and, further, it is not really desirable to spend a great deal of time on devices which, if not yet obsolete, are on the way to being so.

Tetrode and pentode

In many applications of electronics, particularly communications, it is required to operate at very high frequencies. Now because, as you will know, the reactance of a capacitor decreases as the frequency rises, the effect of stray capacitance in a circuit at high frequencies becomes very important. The signal can, in effect, be shorted out by this capacitance.

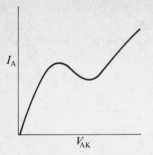

Fig. 9.16. Output characteristics of a tetrode.

There is, of course, capacitance between the three electrodes of the triode and although the values are only in the order of a few picofarads they become very damaging at high frequencies. Perhaps surprisingly, the most damaging by far is the capacitance between the anode and grid. Miller discovered that the capacitance between the input and output of an amplifier *behaves* as though it were a much larger capacitance in parallel with the input and hence tending to short it out (the Miller effect). The input–output capacitance is *amplified* by the amplifier by about the same amount as the input voltage.

To reduce this anode–grid capacitance for high-frequency work a second grid, known as the *screen grid*, is introduced between the grid (now called the *control grid*) and anode. Although the screen grid is connected to a positive potential to accelerate the electrons, most of which will pass through it and travel to the anode, it is earthed for high frequencies by a large capacitor. An earthed plate between the plates of a capacitor (here the anode and grid) will substantially reduce the capacitance.

These valves, called tetrodes, have never been much used because of the presence of a kink in the output characteristics (Fig. 9.16). As the anode voltage is increased, electrons on reaching the anode knock out secondary electrons which are pulled back to the screen grid, the anode current actually decreasing over part of the characteristic as V_{ak} is increased. To overcome this a third grid is inserted between the screen grid and anode and is normally connected, sometimes internally, to the cathode. This grid is called the *suppressor* (it suppresses secondary electrons) and this valve is known as the pentode (Fig. 9.17). The suppressor pushes the secondary electrons back to the anode.

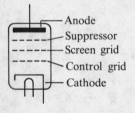

— Anode
— Suppressor
— Screen grid   *Fig. 9.17. Symbol of pentode.*
— Control grid
— Cathode

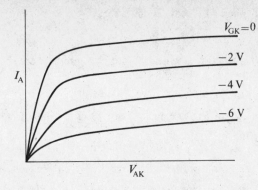

*Fig. 9.18. Output charac-
teristics of a pentode.*

The pentode is used in a similar way to the triode, although its output characteristics are somewhat different (Fig. 9.18). They are, in fact, much more like the output characteristics of a transistor, except for the higher voltages involves, and clearly r_a is considerably higher than for the triode, being usually in the order of 1 MΩ or more. g_m values are similar to the triode because the geometry of the control grid is similar, and hence, since $\mu = r_a g_m$ values of μ will be considerably higher, perhaps as high as 10 000.

Many other valves are manufactured, often for special purposes. It is very common to find two valves, and even three, in the same glass envelope, sometimes sharing the same cathode. Thus double diodes and double triodes are often used and even double pentodes have been manufactured. However, they can be considered as separate valves as far as circuit design and analysis are concerned.

9.B Test questions

1. The main function of the grid in the triode is
    (a) to enable the emitting efficiency of the cathode to be increased
    (b) to control the flow of electrons to the anode
    (c) to enable the triode to pass current in both directions

2. Define the amplification factor, μ, of a triode.

3. Draw a circuit which may be used to find the output and/or the transfer characteristics of a triode.

4. Sketch sets of output and transfer characteristics for a triode, indicating typical values.

5. The anode resistance, or dynamic output resistance of a triode, r_a, is defined as

    (a) $\dfrac{\text{Change in } V_{AK}}{\text{Change in } I_A}$ at constant V_{GK}

    (b) $\dfrac{\text{Change in } I_A}{\text{Change in } V_{AK}}$ at constant V_{GK}

(c) The change in V_{AK} having the same effect as a one volt change in V_{GK}.

(d) The slope of the output characteristics.

6. Estimate r_a for the triode whose output characteristic are shown in Fig. 9.8 at an anode voltage of 200 V and a grid voltage of -2 V.

7. Define the mutual conductance, g_m, of a triode.

8. A triode has a g_m of 5 mA/V and a μ of 120. The value of r_a is
  (a) 600 kΩ
  (b) 24 kΩ
  (c) 24 Ω
  (d) cannot be found from the information given

9. A triode has the following maximum values: $V_{AK} = 500$ V; $I_A = 100$ mA; and maximum power $= 2 \cdot 5$ W. It is operated with a value of V_{AK} of 200 V. What is the maximum permissible I_A?
  (a) $12 \cdot 5$ mA   (c) $1 \cdot 25$ mA
  (b) 100 mA   (d) 50 mA

10. The screen grid in the tetrode is introduced
  (a) to allow the Miller effect to operate
  (b) to take some of the current from the anode and thus reduce the power dissipated at the anode
  (c) to decrease the anode–grid capacitance
  (d) to suppress secondary electrons

11. Explain how it is possible for the screen grid to be both at a high voltage and at earth potential as far as is concerned.

12. The suppressor in a pentode is
  (a) to suppress secondary electrons
  (b) to reduce the capacitance between anode and grid
  (c) to reduce the screen–anode capacitance .
  (d) to remove positive ions from the valve

13. Sketch typical output characteristics for a pentode.

14. Compare the relative values of r_a, g_m, and μ for the triode and pentode.

15. The triode, whose output characteristics are shown in Fig. 9.8 has its grid externally connected to the cathode. It thus behaves as a diode, and as such is connected in series with a 10-kΩ resistor to a 250-V d.c. supply. Draw a load line on the characteristics and estimate the current flowing.

16. Repeat question 15 but with the grid now at -2 V with respect to the cathode.

Questions 17–20 deal with the triode whose output characteristics are given in Fig. 9.11, now operated from a 400-V supply.

17. If the working point Q is to remain the same ($V_{gk} = -1 \cdot 5$ V, $V_{ak} = 185$ V, and $I_a = 1 \cdot 15$ mA) the value of the load resistor is
  (a) 100 kΩ (c) 400 kΩ
  (b) 200 kΩ (d) 150 kΩ

18. For an input of $1 \cdot 5$ V peak, the voltage gain will be
  (a) 57   (c) 115
  (b) 72   (d) 143

19. If the anode load is $100 \text{ k}\Omega$ (still with a 400-V supply) find the new position of Q for a -1.5 V grid bias.

20. With the Q point as in question 19 find the average power delivered by the supply.

21. The triode whose characteristics are given in Fig. 9.11 now has its anode connected directly to a 250-V d.c. supply. Draw the load line and estimate the anode current for a grid bias of -1.5 V.

22. For the conditions of question 21 estimate the peak to peak anode current and voltage for a sinusoidal input voltage of 0.5 V peak.

9.8 The cathode-ray tube (CRT) and its applications

Certain substances fluoresce, or glow, when bombarded with a stream of electrons. The cathode-ray tube (CRT) makes use of this phenomenon: a stream of electrons, in a vacuum, hitting a fluorescent screen which is viewed from the other side. The colour of the fluorescence is a property of the actual substance used to coat the screen, as is the length of time for which the glow remains after bombardment. These factors will be discussed further when the applications of the tube are dealt with.

Apart from the fluorescent screen, the main parts of a CRT are the electron gun and the deflection system, the former to produce a beam of electrons and to focus it on the screen and the latter moving the spot so produced to any desired part of the screen.

Electron gun

The electrons are produced by an indirectly heated, oxide-coated cathode (see Fig. 9.19). The grid, which, as in a valve, controls the flow of electrons and hence the brightness of the spot, consists of a tube round the cathode. The electrons are accelerated towards an anode (A1) which has in it a hole through which most of them pass, entering a tube (A2) which is the focusing electrode, and finally passing through another anode (A3) at an even higher positive potential. This system acts rather like an optical lens, and, with suitable voltages applied, will focus the beam of electrons into a spot on the screen, Fig. 9.19.

The third anode (A3) may be several thousands of volts positive with respect to the cathode, although for reasons which will appear later, it is often connected to earth, the cathode being made negative with respect to it. The A2 anode is taken to a somewhat lower and variable voltage, the focus control, and finally the grid is connected to a variable voltage a little negative with respect to the cathode, the brightness control.

Deflection system

There are two ways of deflecting the beam of electrons, either by an *electrostatic* or *magnetic* field. In either case it must be made possible to

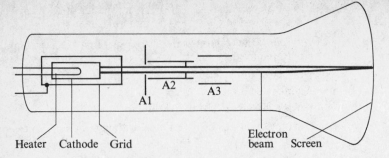

Heater   Cathode   Grid

Electron
beam   Screen

Fig. 9.19. CRT electron gun and focusing system.

deflect the spot both horizontally (X deflection) or vertically (Y deflection).

Electrostatic deflection is produced by using two pairs of plates arranged as shown in Fig. 9.20. In most CRTs one of each pair is earthed through connection to the final anode (A3), the deflecting voltage being applied to the other plate of the pair.

If Y_2 is earthed a positive voltage applied to Y_1 will attract the electrons and the spot will move upwards. Similarly, a positive voltage on X_1 with X_2 earthed will pull the beam towards the reader producing a deflection on the screen to the left as viewed from the front.

It can be shown that as long as the deflection is not too large, the spot deflection will be proportional to the deflecting voltage, a very useful property.

Electrostatic deflection is not, however, very convenient for one very important application of the CRT, the television tube, because it leads to a rather long tube of not very large screen area. Modern picture tubes are required to be of large screen area but short in length, requiring a large angle of deflection. This would mean deflecting voltages which are far too large to be convenient, and as a consequence such tubes employ magnetic deflection. This type of deflection depends on the motor principle, namely that a force is exerted on a current-carrying conductor in a magnetic field. The electron beam is a current, and will be deflected by a magnetic field.

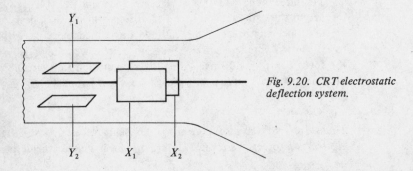

Fig. 9.20. CRT electrostatic deflection system.

Coils are placed outside the tube and the spot is deflected by an amount proportional to the current in the coils.

The electrostatic CRT is normally used in laboratory instruments, not only because it is generally more useful to have a deflection proportional to voltage but also because it is a more accurate instrument.

The cathode-ray oscilloscope (CRO)

Fig. 9.21 shows an electrostatic CRT used in an oscilloscope connected to its power supply (note the circuit symbol for a CRT). The power supply consists of a chain of resistors across a large voltage (EHT).

It will be seen that X_2 and Y_2 are earthed, the inputs being taken to X_1 and Y_1. This enables the inputs to be applied with respect to earth and, as it is desirable for the deflecting voltages to be in the same order of magnitude as the final anode voltage, it is necessary to earth the final anode, that is the positive side of the EHT supply.

There are, in fact, two inputs to both X_1 and Y_1. One of the inputs deflects the spot by an amount proportional to the signal voltage and the other is a d.c. shift to enable the spot to be positioned where desired on the screen. You will become familiar with all the controls shown during your practical work in the laboratory.

One of the main applications of the CRO is to observe the shape of a waveform, that is a graph of the waveform against time. For this purpose

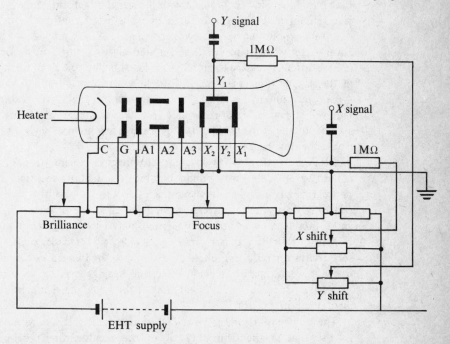

Fig. 9.21.

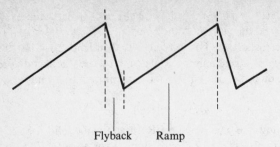

Fig. 9.22. *Time-base waveform.*

a CRO incorporates a *time base*, an electronic circuit producing a ramp, or a voltage rising at a constant rate. This is applied to the X_1 plate and moves the spot horizontally from left to right at a constant speed, rapidly returning at the end of the trace (the flyback). The waveform required to do this is shown in Fig. 9.22.

It is not difficult to make the time-base circuit also produce a negative pulse during the flyback period. This pulse, called a blanking pulse, is applied to the grid, cutting off the electron stream, or blanking out the spot, during flyback.

If the signal to be observed is now applied to the Y_1 plate, a plot of the applied voltage against time will be seen. It is, of course, necessary for the time base to run in synchronism with the signal waveform, and a synchronizing circuit is used to ensure that the time-base waveform occupies just one, or a whole number of, cycles of the signal, so that successive traces cover exactly the same path and a stationary picture can be observed.

It is often necessary to amplify the signal being studied and this is done in the Y amplifier (Fig. 9.23). This shows two switches. S1 enables the synchronization to be taken either from the Y signal or, as is sometimes required, from an external source. S2 allows an external signal to be applied to the X plates because there are a few applications of the CRO where the time base is not required.

CRTs are available with different phosphors producing traces of various colours, green being very common. Blue is suitable if the trace is to be photographed. For low-frequency applications, where the spot moves relatively slowly, a long afterglow, perhaps of several seconds, is required, and is produced by a suitable fluorescent material.

Many CROs have two Y inputs and display two traces at once, which is very useful for observing phase relationships. The two traces are produced either by having two separate electron beams each with its own Y deflection, or by high speed switching of the two signals, in turn, to the Y plate of a normal CRT.

The only way to get to know what is, and is not, possible with a CRO is to use one in the laboratory. Read the manufacturer's instructions, particularly with regard to synchronizing the time base. The CRO is an

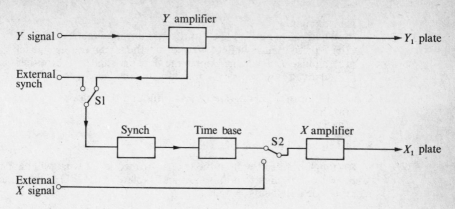

Fig. 9.23.

extremely powerful tool in the laboratory. It is hard, nowadays, to imagine an electronics laboratory without them, although even as recently as just before the Second World War they were relatively scarce.

Other applications

There is no need to explain that the CRT is also used to display a picture. In a TV receiver there are two time bases, one moving the spot downwards and the other, at a higher frequency, moving it from left to right, thus producing a raster of lines on the face of the tube. The picture is formed by varying the brightness of the spot by applying the signal, suitably amplified, to the grid.

Other applications, some of which you might meet in your course, include radar and computers.

9.C Test questions

1. Describe, with the aid of a diagram, the electron gun of a CRT.

2. Explain why the final anode is usually earthed in a CRT.

3. Magnetic deflection is used in TV picture tubes
   (a) because the coils occupy less space than the plates used for electrostatic deflection
   (b) to make possible the very wide angle deflection required
   (c) because of the high scanning frequency
   (d) because better picture quality is possible

4. Electrostatic deflection is usually used in tubes in CROs because
   (a) greater accuracy is possible for the same cost
   (b) long thin tubes are desirable
   (c) a long afterglow is needed
   (d) the coils might pickup interference

5. Draw a circuit diagram of a power supply operating a CRT including the X and Y shifts.

6. A blanking pulse is
   (a) a negative pulse applied to the grid during flyback
   (b) a pulse used to deflect the spot off the screen during flyback
   (c) a pulse used to synchronize the time base and may be either
       derived from the signal or externally applied

7. It is not possible to operate a CRO without a time base.
   (a) true
   (b) false

8. List three applications of a CRT.

9. Although not dealt with in the text, try to work out what will happen
if a sine wave is applied to both the X- and Y-plates of a CRO at the same
time, the time base being switched off.

10. Repeat question 9 with identical sine waves on the X- and Y-plates
except for a 90° phase difference between them.

Solutions to test questions

2.A   1. c.   2. a.   3. b.   4. a.   5. c.   6. a.   7. a.   8. b.   9. c.
10. b.   11. c.   12. b.   13. a.   14. b.   15. a.   16. b.   17. c.
18. b.   19. a.   20. b.   21. b.   22. a.   23. c.   24. b.

2.B   1. c.   2. b.   3. c.   4. See Fig. 2.18 and Fig. 2.19.   5. a.
6. a.   7. a.   8. c.   9. See Fig. 2.20.
10. See Fig. 2.21. Intercept is I_{CB0}.   11. c.   12. c.   13. b.   14. b.
15. a.   16. b.
17. c; Current gain = 0·995, voltage gain = 0·995 × (1000/10) = 99·5, and power gain = 0·995 × 99·5 = 99.
18. a.   19. a.   20. See Figs. 2.24, 2.26, 2.27.
21. c; This is eqn (2.8) transposed.
22. d; α falls from 0·997 to 0·993.
23. a; $I_{CB0}/(1 - \alpha) = I_{CB0} (1 + \beta)$.
24. c.
25. d; 1 V at base would produce 1 V/1 kΩ = 1 mA of I_B hence 199 mA (β) of I_C, giving output voltage 199 × 2 = 398 or 400. (gain)
26. a; Voltage gain × current gain = 400 × 200 = 80 000.
27. b.
28. c; It can never quite equal unity as there must always be some I_B. Answer d might imply a gain greater than unity.
29. 200   30. 1 kΩ.

2.C   1. b; (input resistance is very high).
2. b; (carriers are holes attracted to the negative drain).
3. b.   4. a; (hence the high input resistance).   5. a.
6. (i) a, (ii) b, (iii) c.   7. b.   8. 0·4 mA/V.   9. f.   10. b.
11. (i) steep, (ii) horizontal.   12. c.   13. b.   14. b.   15. b.

3.A   1. See Figs. 3.1 and 3.2.   2. a.   3. a.   4. d.
5. a; 300 V r.m.s. = 424 V peak. Av. output = 2 × 424/π = 270 V. 270 V/135 kΩ = 2 mA.
6. b.   7. b.   8. See text.   9. See text.   10. See Text.
11. See Fig. 3.1.   12. a.   13. a; 0·005/0·002 = 2·5 V.   14. d.
15. c; (b is the peak to peak amplitude).   16. b.   17. b.   18. b.
19. a.   20. c.
21. 9·7%. At 1 A output falls 5 V so that V_{AV} is 25·8 V. Hence reg = (28·3 − 25·8)/25·8 × 100 = 9·7%.
22. a; Reactance of coil at 100 Hz = 628 Ω. Reactance of C at 100 Hz = 1·59 Ω

$$\text{Ripple} = \frac{1·59}{628 - 1·59} \times 1·25 = 3·2 \, \text{mV}.$$

3.B   1. d; voltage across R = 13 V. b and c would give total currents of less than 30 mA and cannot be used. a gives a total current of 39 mA and d of 33·3 mA, hence we might as well use d.
2. b; total current is 33·3 mA hence if load is 30 mA. I_Z must be 3·3 mA.

3. a; max I_Z is $33\cdot3$ mA when I_L is zero.
$33\cdot3 \times 12 = 400$ mW.
4. I_Z max is $0\cdot5/12 = 0\cdot0417$ A or $41\cdot7$ mA.
Voltage across R (390 Ω) is $(0\cdot39 \times 41\cdot7) = 16\cdot25$ V.
$V_{IN} = 16\cdot25 + 12 = 28\cdot25$ V.
Minimum is just about zero volts across R.
Range is 12–28·25 V.
5. If $I_Z = 41\cdot7$ mA the total current is $71\cdot7$ mA and the voltage across
$R = 71\cdot7 \times 0\cdot39 = 28$ V; $V_{IN} = 40$ V. If I_Z is zero the total current is
30 mA and the voltage across R is $30 \times 0\cdot39 = 11\cdot7$ V; $V_{IN} = 23\cdot7$ V.
Therefore the range is 23·7–40 V.

3.C   **1.** b.
2. b; the true value is zero, but the average is often defined as the average
of a half cycle, in which case the answer would be c.
3. d; the peak current is 1 mA and the half-wave rectified average, which
will be read on the meter, will be $1/\pi$ or $0\cdot318$ mA.
4. Because it would completely alter the current we were measuring by
half-wave rectifying it.
5. See Fig. 3.17.
6. c; 10 A r.m.s., 14·14 A peak, and the full-wave rectified average of this
is 9 A. There is no mention in the question of the meter being recalibrated to
to read r.m.s. values.
7. d; it will read the average value $\times 1\cdot11$ but there is no way of finding
the average value in this question.
8. b; form factor is the ratio of r.m.s. to average values. Thus the average
value in this case is $100/1\cdot01 = 99$ V and the meter will read $99 \times 1\cdot11 =$
$109\cdot9$ V.
9. (a) Centre junction. (b) Two outside junctions.
10. Cathode-gate SCR.   **11.** Into the gate.   **12.** See Fig. 3.22.
13. The minimum would be zero and the maximum half a cycle.

4.A   **1.** See text.   **2.** a.   **3.** b.   **4.** See text.

5. b; $\dfrac{2 \text{ mV} \times 0\cdot5}{(1\cdot5 + 0\cdot5)}$ **6.** a; 10 mA in 0·5 Ω.

7. a; output resistance of amplifier equals the meter resistance.
8. d; 5 mV/0·5 mV.
9. d; current in meter = 10 mA
current at input = $0\cdot5$ mV/$0\cdot5$ MΩ = 10^{-9} A
current gain = $10 \times 10^{-3}/10^{-9} = 10^7$.
10. a; $A_i \times A_v = 10^7 \times 10 = 10^8$ or 80 dB.
11. a; 5 mV/2 mV = 2·5.
12. c; power in load = $10 . 10^{-3} \times 5 . 10^{-3} = 5 . 10^{-5}$ W
power from accelerometer
$\qquad 2 . 10^{-3} . 10^{-9} = 2 . 10^{-12}$ W
Gain $5 . 10^{-5}/2 . 10^{-12} = 2\cdot5 \times 10^7$ or 74 dB.
13. See text.   **14.** See text (a) no, (b) yes.   **15.** See text.
16. See text.   **17.** See text.   **18.** c.   **19.** b.   **20.** a.
21. c; voltage ratio is $10/10^{-3} = 10^4$, a power ratio of 10^8 or 80 dB.
22. a; 40 dB is 10 000 : 1. Hence output is 1 W.
23. d; 50 dB is 100 000 : 1. Hence output noise is 10 W/100 000 = 10^{-4} W.
This is a voltage of $\sqrt{(10^{-4} \times 15)} = 39$ mV.

4.B

1. An amplifier working on the linear part of its characteristics.
2. Intercepts at $10\,\text{V}$ and $40\,\text{mA}$.
3. $100\,\mu\text{A}$; it cannot be less, and if more the signal will drive into the less linear regions. Also a larger bias will waste power.
4. $17.5\,\text{mA}$ and $5.6\,\text{V}$.　　**5.** c; $(10-0.6)\,\text{V}/100\,\mu\text{A} = 94\,\text{k}\Omega$.
6. b; about $10\,\text{V}/100\,\mu\text{A} = 100\,\text{k}\Omega$.
7. c; current swings from 4.5 to $30\,\text{mA} = 25.5\,\text{mA}$ peak to peak. $25.5/(2\sqrt{2}) = 9\,\text{mA}$.
8. b; $25.5\,\text{mA}/200\,\mu\text{A} = 127.5$.
9. d; voltage swings from 9 to $2.5 = 6.5\,\text{V}$ peak to peak. $6.5/(2\sqrt{2}) = 2.3\,\text{V}$.
10. d; $9 \times 10^{-3} \times 2.3 = 20.7 \times 10^{-3}$.
11. b; $v_{\text{in}} = 4.5 \times 10^{3} \times 0.1 \times 10^{-3} = 0.45\,\text{V}$ peak. Gain $= 6.5/0.9 = 7.2$.

12. d; input power $= \dfrac{0.45}{2} \times \dfrac{0.1 \times 10^{-3}}{2} = 22.5 \times 10^{-6}$.

Power gain $= 20.7 \times 10^{-3}/22.5 \times 10^{-6} = 920$
$$= 29.6\,\text{dB}.$$
13. a; input current $= 100\,\mu\text{A}$ peak,
input current $= 100\,\mu\text{A} \times 0.5\,\text{k}\Omega = 50 \times 10^{-3}\,\text{V}$ peak, therefore

Input power $= \dfrac{100 \times 10^{-6} \times 50 \times 10^{-3}}{\sqrt{2} \times \sqrt{2}} = 2.5 \times 10^{-6}\,\text{W}$.

Gain $= 20.7 \times 10^{-3}/2.5 \times 10^{-6} = 8280$ or $39.2\,\text{dB}$.
Not reliable because input voltage will not be sinusoidal.
14. a; $10\,\text{V} \times 17.5\,\text{mA} = 175\,\text{mW}$.　　**15.** c; $17.5\,\text{mA} \times 5.6\,\text{V} = 98\,\text{mW}$.

4.C

1. See text.

2. b; $\beta = \dfrac{0.992}{1-0.992} = 124$, $I_{\text{C}} = 124 \times 10\,\mu\text{A} = 1.24\,\text{mA}$.

3. a; $\beta = 199$.　　**4.** b; $V_{\text{E}} = 2\,\text{V}$.　　**5.** d; $2.6\,\text{V}/0.1\,\text{mA} = 26\,\text{k}\Omega$.
6. b; $I_{\text{B}} = 1\,\text{mA}/100 = 10\,\mu\text{A}$.　　**7.** b; $(15-2.6)\,\text{V}/0.11\,\text{mA} = 112.7\,\text{k}\Omega$.
8. a; $15 - (5 \times 1) = 10\,\text{V}$, $V_{\text{CE}} = 10 - 2 = 8\,\text{V}$.

9. c; $C = \dfrac{1}{2\pi\,100 \times 200}\,\text{F} = 7.95\,\mu\text{F}$.　　**10.** b.

11. a; $V_{\text{CE}} = 10 - (1 \times 5) = 5\,\text{V}$.
$R_{\text{B}} = (5-0.6)/0.05 = 88\,\text{k}\Omega$.
12. V_{CE} falls, I_{B} falls, I_{C} falls. Yes.

4.D

1. Intercepts $20\,\text{V}$ and $2.67\,\text{mA}$.　　**2.** $1\,\text{mA}$; $12.6\,\text{V}$.
3. $0.4\,\text{mA}$ to $1.7\,\text{mA}$.　　**4.** $(1.7 - 0.4)/(2\sqrt{2}) = 0.46\,\text{mA}$.
5. Swing 16.8 to $7.2\,\text{V}$ peak to peak $= 9.6\,\text{V}$.　　**6.** $9.6/(2\sqrt{2}) = 3.4\,\text{V}$.
7. $9.6/4 = 2.4$.　　**8.** $3.4 \times 0.46 = 1.6\,\text{mW}$.　　**9.** $20 \times 1 = 20\,\text{mW}$.
10. $4/1 = 4\,\text{k}\Omega$.
11. Gate voltage $1/20 \times 20 = 1\,\text{V}$, so source must be at $5\,\text{V}$, $5\,\text{k}\Omega$ needed

4.E

1. a; $i_{\text{b}} = \dfrac{v_{\text{in}}}{6\,\text{k}\Omega}$, current in load $i_{\text{b}} \times 120 = 20\,v_{\text{in}}$. $v_{\text{o}} = 20\,v_{\text{in}} \times R_{\text{L}}$.

Gain $= 20\,R_{\text{L}} = 100$, $R_{\text{L}} = 5\,\text{k}\Omega$.
2. b; R_{L} and h_{oe} in parallel must be $5\,\text{k}\Omega$, h_{oe} is equivalent to $1/100\,\text{M}\Omega$ or $10\,\text{k}\Omega$, R_{L} must be $10\,\text{k}\Omega$.

3. a;. transistor produces gain of 120 only half of which will flow in R_L.
4. c: $60 \times 100 = 6000$ or $37 \cdot 8$ dB.

5. b; $\beta = \dfrac{0 \cdot 996}{1 - 0 \cdot 996} = 249$, h_{oe} is equivalent to $10 \, k\Omega$, $R_L = 2 \cdot 5 \, k\Omega$

hence fraction in R_L is $10/12 \cdot 5$ or $0 \cdot 8$. Gain $= 249 \times 0 \cdot 8 = 199$.
6. c; $R_{IN} = h_{ie} + (1 + h_{fe}) R_L$.   7. a; $(1 + h_{fe})$.   8. b; see eqn (4.5).
9. b; $0 \cdot 996 \times 400$.
10. d; load $= 20 \, k\Omega$ in parallel with $5 \, k\Omega = 4 \, k\Omega$. Gain $= 4 \times 3 = 12$.
11. b; $\mu = g_m \times r_a = 3 \times 20 = 60$.   12. See text.
13. $\mu R_L / (r_a + R_L) = (60 \times 5)/25 = 12$.

5.A

1. See text.   2. See text.   3. See text.   4. Cuts axes at $10 \, V$ and $4 \, mA$.
5. $(10 - 0 \cdot 6)/0 \cdot 01 = 940 \, k\Omega$.   6. $(10 - 0 \cdot 6)/0 \cdot 05 = 188 \, k\Omega$.
7. c; (actually $1 \cdot 996 \, k\Omega$).   8. a; (actually $0 \cdot 998$).   9. a.
10. Cuts voltage axis at $6 \cdot 25 \, V$.   11. 185.   12. d.   13. a.   14. a.
15. Cuts axes at $10 \, V$ and $10 \, mA$.   16. Cuts voltage axis at $7 \, V$.
17. 128.   18. 64.   19. $64 \times 141 = 9024$.
20. $v_o = 6 \cdot 1 - 4 = 2 \cdot 1 \, V$ peak to peak.
$v_{in} = 12 \, k\Omega \times 0 \cdot 02 \, mA = 0 \cdot 24 \, V$ peak to peak.
$A_v = 2 \cdot 1/0 \cdot 24 = 8 \cdot 75$.
21. $v_o = 5 \cdot 4 - 2 \cdot 2 = 3 \cdot 2 \, V$ peak to peak.
$v_{in} = 50 \, \mu A \times 0 \cdot 8 \, k\Omega = 0 \cdot 04 \, V$ peak to peak.
$A_v = 3 \cdot 2/0 \cdot 04 = 80$.
22. $8 \cdot 75 \times 80 = 700$.   23. $700 \times 9024 = 6 \, 316 \, 800$ or 68 dB.
24. T_1 $190 \times 0 \cdot 76 = 144 \cdot 4$.
T_2 $130 \times 0 \cdot 5 = 65$.
Total, $144 \cdot 4 \times 65 = 9386$.
25. T_1 $v_o = 190 \, i_{IN} \times 0 \cdot 604 = 114 \cdot 8 \, i_{IN}$
     $v_{in} = 12 \, i_{IN}$
     $A_v = 114 \cdot 8/12 = 9 \cdot 6$
  T_2 $v_o = 130 \times 0 \cdot 5 \, i_b = 65 \, i_b$
     $v_{in} = i_b \times 0 \cdot 8$
      $A_v = 65/0 \cdot 8 = 81 \cdot 3$
Total $A_v = 9 \cdot 6 \times 81 \cdot 3 = 780$.
26. $9386 \times 81 \cdot 3 = 7 \, 321 \, 080$ or $68 \cdot 6$ dB.   27. See text.
28. $8^2 \times 100 = 6400 \, \Omega$.   29. b; $50 \times \sqrt{1000}/10 = 500$.   30. See text.
30. See text.   31. See text.   32. $250 + 200 \times 251 = 50450$ $\alpha = 0 \cdot 9999$.
33. d; total current $= (15 - 0 \cdot 6)/10 = 1 \cdot 44 \, mA$.
34. a; $(15 - 5)/0 \cdot 72 = 14 \, k\Omega$.   35. b.   36. d.
37. b; $0 \cdot 25 \times 14 \, k\Omega = 3 \cdot 5 \, V$ i.e. T_1 falls $3 \cdot 5$ and T_2 rises $3 \cdot 5$ hence change
is $7 \, V$ with T_2 more positive.
38. See text.   39. a.   40. b.   41. b.   42. See text.
43. c; half power is a fall of 3 dB.   44. a; half power.   45. See text.

5.B

1. See text.

2. d; $C = \dfrac{1}{4\pi^2 f^2 L} = \dfrac{1}{4\pi^2 . 10^{12} \, 100 . 10^{-6}} \, F_2 = 253 \, pF$.

3. a; $\dfrac{L}{CR} = \dfrac{100 . 10^{-6}}{253 . 10^{-12} . 5} \, \Omega = 79 \, k\Omega$.

4. a; $Q_o = \dfrac{\omega L}{R} = \dfrac{2 . \pi . 10^6 . 100 . 10^{-6}}{5} = 125 \cdot 7$.

5. b; bandwidth $= \dfrac{f_0}{Q_0} = \dfrac{10^6}{126}$ Hz $= 7.94$ kHz.

6. a; $A_v = \dfrac{50}{\sqrt{2}} = 35.35$.

7. 1 MHz $- 7.9$ kHz $= 992.1$ kHz.

8. Q_0 becomes a quarter of its value (i.e. 31.5) hence R goes up 4 times to $20\,\Omega$.

9. Impedence is divided by 4 hence the gain is divided by 4.

10. See text.

5.C

1. See text.   **2.** 1.1 A $\times 7.5$ V $= 8.25$ W.

3. $n^2 = 4$ hence resistance seen by transformer $= 10\,\Omega$.

$10\,\Omega$ at 1.1 A $= 11$ V,

hence a.c. load line cuts voltage axis at $7.5 + 11 = 18.5$ V.

4. Voltage swing $17.5 - 1.3 = 16.2$ V,

current swing $1.75 - 0.1 = 1.65$ A.

5. $\dfrac{16.2 \times 1.65}{8} = 13.4$ W.   **6.** $\dfrac{6.9}{15} = 0.46$ kΩ.

7–10. See text.

6.A

1. See text.   **2.** a.   **3.** b.   **4.** b.   **5.** c.   **6.** d.   **7.** c.

8. d; $A_v = -10\,000$

$\qquad \beta = 0.02$

$\quad A_v\beta = -10\,000 \times 0.02$

$\qquad\quad = -200$

$1 - A_v\beta = 1 + 200 = 201$

$\qquad A_{vf} = \dfrac{-10\,000}{201} = -49.75$.

9. b; loop gain $= A_v\beta$.   **10.** a; feedback factor $1 - A_v\beta$.

11. a; $20 \log_{10} \dfrac{1}{1 - A_v\beta} = 20 \log \dfrac{1}{201}$

$\qquad = 20 \times (-2.3) = -46$ dB.

12. b; A_v falls to $0.8 \times (-10\,000) = -8000$

$\quad A_v\beta = -8000 \times 0.02 = -160$

$1 - A_v\beta = 161$

$\qquad A_{vf} = \dfrac{-8000}{161} = -49.69$.

$\qquad$ fall $= \dfrac{49.75 - 49.69}{49.75} \times 100 = 0.12$ per cent.

13. d; $f_1' = \dfrac{f_1}{1 - A_v\beta} = \dfrac{25}{201} = 0.12$ Hz.

14. b; $f_2' = f_2(1 - A_v\beta) = 50 \times 201 = 10\,050$ kHz

$\qquad\qquad\qquad\qquad\qquad\qquad\quad = 10.05$ MHz.

15. b; $\dfrac{5\%}{1 - A_v \beta} = \dfrac{5}{201} = 0.025$ per cent.

16. Let gain without feedback (nominal) be A_{v1}.
Let gain without feedback when fallen 35 per cent be A_{v2}.
Let gain with feedback originally be A_{vf1}.
Let gain with feedback in the second case be A_{vf2}.

$$A_{vf1} = \frac{A_{v1}}{1 - A_{v1}\beta} = \frac{-5000}{1 + 5000\beta}$$

$$A_{vf2} = \frac{A_{v2}}{1 - A_{v2}\beta} = \frac{-5000(1 - 0.35)}{1 + A_{v2}\beta} = \frac{-3250}{1 + 3250\beta}$$

Now $A_{vf2} = A_{vf1}(1 - 0.01) = 0.99 A_{vf1}$.

Hence, cross multiplying

$$A_{vf1}(1 + 5000\beta) = -5000$$
$$0.99 A_{vf1}(1 + 3250\beta) = -3250$$

and dividing

$$\frac{(1 + 5000\beta)}{0.99(1 + 3250\beta)} = \frac{5000}{3250} = 1.538.$$

$$(1 + 5000\beta) = 1.538 \times 0.99(1 + 3250\beta)$$
$$= 1.523 + 4949\beta$$
$$51\beta = 0.523$$

$$\beta = \frac{0.523}{51} = 0.01$$

$$A_{vf1} = \frac{-5000}{1 + 5000 \times 0.01} = \frac{-5000}{51} = -98.$$

17. $A_{vf} = \dfrac{-1\,000\,000}{1 + 100\,000} = \dfrac{1\,000\,000}{100\,001} = 9.9999.$

18. Oscillates if $A_v \beta = 1$

Hence $A_v = \dfrac{1}{\beta} = \dfrac{1}{0.1} = 10.$

19. $A_{vf} = \dfrac{A_v}{1 - A_v \beta}$ for positive or negative feedback.

Hence $A_{vf} = \dfrac{100}{1 - 0.8} = \dfrac{100}{0.2} = 500.$

20. A_v becomes 50

$$A_{vf} = \frac{50}{1 - 0.4} = \frac{50}{0.6} \doteqdot 83.3$$

percentage fall $= \dfrac{500 - 83.3}{500} \times 100 = 83$ per cent.

21. Oscillates if $A_v \beta = 1$.

Hence $A_v = \dfrac{1}{0.008} = 125.$

6.B   1. Current.   2. Series.   3. High.   4. High.   5. Voltage.
6. Shunt.   7. a.
8. 1. As the voltage gain from base to emitter is about one, and as the two output loads are equal, the voltage across the collector load must be (about) equal to that across the emitter load.
9. For v_{o1} it is current, but for v_{o2} voltage feedback.
10. Following from the answers to question 9 output one will be high resistance and output two low resistance.
11. They would be $180°$ out of phase, with v_{o2} in phase with the input voltage.
12. b.   13. d. It is equal to 1 MΩ in parallel with 100 kΩ.

14. $-6\,\text{V}$; $v_o = -1\left(\dfrac{2}{1} + \dfrac{1}{0\cdot25}\right) = -6\,\text{V}$.

15. $-10\,\text{V}$; $v_o = -500\left(\dfrac{1}{100} + \dfrac{-1}{100} + \dfrac{2}{100}\right) = -5 + 5 - 10$.

16. Gain of Fig. 6.26 $= -2$, i.e. the magnitude is 2.

100 dB is a gain of antilog $\dfrac{100}{20}$ = antilog $5 = 100\,000$.

Gain bandwidth product $= 100\,000 \times 5 = 0\cdot5\,\text{MHz}$

Bandwidth $= \dfrac{0\cdot5}{2}\,\text{MHz} = 250\,\text{kHz}$.

17. Gain bandwidth product $= 0\cdot5\,\text{MHz}$;

Gain $= \dfrac{0\cdot5}{1} = 0\cdot5\,\text{MHz}$.

18. See text.
19. See text.
One way of making the output variable is to take the base of T_2 (Fig. 6.19) to a variable resistor across the output. Another is to take the emitter of T_2 to a potentiometer across the zener, effectively altering the reference voltage.
20. $R_{in} = h_{ie} + (1 + h_{fe})\,R$
  $= 1\cdot5 + (1 + 200) \times 1$
  $= 1\cdot5 + 201$
  $= 202\cdot5\,\text{k}\Omega$

Approximate gain $= \dfrac{R_L}{R_E} = \dfrac{5}{1} = 5\,(180°\text{ phase shift})$.

21. $A_v = \dfrac{-200 \times 5}{1\cdot5} = -666\cdot7$ $\beta = \dfrac{1}{5} = 0\cdot2$

  $A_{vf} = \dfrac{-666\cdot7}{1 + 666\cdot7 \times 0\cdot2} = \dfrac{-666\cdot7}{134\cdot3} = -4\cdot96$.

22. R_{in} becomes $202\cdot5\,\text{k}\Omega$ in parallel with 84 kΩ and 16 kΩ in parallel. This gives $R_{in} = 12\cdot6\,\text{k}\Omega$.

7.A   1. b.   2. a.   3. a.   4, 5, 6, 7. See text.
8. b; period 1/50 000s $= 20\,\mu s$ Pulse $= 2\,\mu s$ space $= 18\,\mu s$.
9. a.   10. b; at f_2 the reactance equals the resistance.
11. c; $\sqrt{(R^2 + X_c^2)}$.   12. a; 1/1·414.   13. c; $0\cdot707 \times 1$.

14. b; f_2 is the upper 3 dB frequency.
15. b; $45°$ at f_2 and leads because it is a capacitive circuit.
16. c; voltage across C $90°$ behind current.   **17.** b; from 15 and 16.
18. d; as 1 MHz is well above f_2 we can assume that a doubling in f gives a halving of the output.
19. a; f is well below f_2 and it can be assumed that output is constant.
20. a; integrates for pulses which are short compared to RC.
21. b; $RC = 10^6 \times 100 \times 10^{-12} = 10^{-4} = 100\,\mu s$.
22. c; differentiates for long pulse.

7.B

1, 2, 3. See text.

4. $L = 41\,\mu H$ $f = \dfrac{1}{2\pi\,41 \times 10^{-6} \times 100 \times 10^{-12}} = 2.5\,MHz$.

Min $h_{fe} = 40/1 = 40$.

5. Min $h_{fe} = 30 = C_1/C_2 = C_1/50$ $C = 1500\,pF$

$C = \dfrac{C_1 C_2}{C_1 + C_2} = \dfrac{50 \times 1500}{50 + 1500} = 48.4\,pF$.

$L = \dfrac{1}{4\pi^2 f^2 C} = \dfrac{1}{4\pi^2\,10^{12}\,48.4 \times 10^{-12}} = 523.4\,\mu H$.

6. Because it still produces $180°$ shift.   **7.** See text.
8. It would still oscillate (at a frequency where the phase shift is $180°$) but the minimum gain would be different.
9. The output would be zero.
10. With $C = 1\,\mu F$ $f = 1/(2\pi\sqrt{6}\,2000 \times 10^{-6}) = 32.5\,Hz$.
With $C = 0.01\,\mu F$ $f = 1/(2\pi\sqrt{6} \times 2000 \times 0.01 \times 10^{-6}) = 3250\,Hz$.
Range 32.5–3250 Hz.

11. $L = \dfrac{1}{4\pi^2 f^2 C} = \dfrac{1}{4\pi^2\,3250^2 \times 0.01 \times 10^{-6}} = 240\,mH$.

12. Lowest $f = \dfrac{1}{2\pi\sqrt{240 \times 10^{-3} \times 10^{-6}}} = 325\,Hz$.

Range 325–3250 Hz.

13. Lowest $R = \dfrac{1}{2\pi\,1000 \times 1 \times 10^{-6}} = 159\,\Omega$.

Highest $R = \dfrac{1}{2\pi\,1 \times 1 \times 10^{-6}} = 159\,000 = 159\,k\Omega$.

Range 159 Ω–159 kΩ.

14, 15, 16. See text.

7.C

1. See text.
2. As $R_3 = R_4$ $C_1 = 5C_2$ (or $C_2 = 5C_1$)
p.r.f. 500 kHz hence $T = 2\,\mu s$
$T = 0.7\,(100 \times 10^3 C_1 + 100 \times 10^3 \times 5C_1)$
$= 0.7 \times 600 \times 10^3 C_1 = 2 \times 10^{-6}$

$C_1 = \dfrac{2 \times 10^{-6}}{0.7 \times 600 \times 10^3} = 4.76\,pF$

$C_2 = 23.8\,pF$.

3. Period of synchronizing pulses $= 1/10^5$ s $= 10\,\mu$s;
130 kΩ produces 110 kHz or 9-μs period; 150 kΩ produces 95 kHz or
150 kΩ produces 95 kHz or 10·5-μs period;
longer period must be chosen hence use 150 kΩ.
4. See text.
5. $RC = 10^5 \times 50 \times 10^{-12} = 50 \times 10^{-7}$
$0{\cdot}7RC = 35 \times 10^{-7}$ s or 3·5 μs.
6, 7. See text.

8.A

1. See text. **2.** See Fig. 8.55. **3.** See Fig. 8.56.
4. It is called an 'exclusive OR' because it excludes the case when we
have A *and* B. The normal logic 'OR' includes the case when we have A
and B and is sometimes called an 'inclusive OR'.
5. See text. **6.** See Fig. 8.57.
7. They are the same. In other words

$$A + A\,.\,B = A.$$

A	B	C	D	Z
0	0	0	0	0
0	0	0	1	1
0	0	1	0	1
0	0	1	1	1
0	1	0	0	1
0	1	0	1	1
0	1	1	0	1
0	1	1	1	1
1	0	0	0	1
1	0	0	1	1
1	0	1	0	1
1	0	1	1	1
1	1	0	0	1
1	1	0	1	1
1	1	1	0	1
1	1	1	1	1

Fig. 8.55.

A	B	$\bar{A}$	$\bar{B}$	$A\bar{B}$	$\bar{A}B$	Z
0	0	1	1	0	0	0
0	1	1	0	0	1	1
1	0	0	1	1	0	1
1	1	0	0	0	0	0

Fig. 8.56.

A	B	AB	Z
0	0	0	0
0	1	0	0
1	0	0	1
1	1	1	1

Fig. 8.57.

A	B	$\overline{A}$	$\overline{A}B$	Z
0	0	1	0	0
0	1	1	1	1
1	0	0	0	1
1	1	0	0	1

Fig. 8.58.

8. See Fig. 8.58.
9. They are the same. In other words
$$A + \overline{A}.B = A + B.$$

10. See Fig. 8.59.　　**11.** See Fig. 8.60.　　**12.** See Fig. 8.61.
13. See Fig. 8.62. The expressions of questions 12 and 13 are two very
important theorems of Boolean algebra known as de Morgan's theorems.
14. $Z = A.\overline{B} + B.C.$ See Fig. 8.63.
15. $Z = A.B + \overline{A}.\overline{B}.$ See Fig. 8.64.

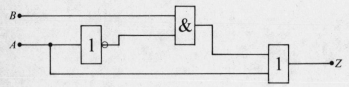

Fig. 8.59.

A	B	C	B+C	A(B+C)		A	B	C	AB	AC	AB+AC
0	0	0	0	0		0	0	0	0	0	0
0	0	1	1	0		0	0	1	0	0	0
0	1	0	1	0		0	1	0	0	0	0
0	1	1	1	0		0	1	1	0	0	0
1	0	0	0	0		1	0	0	0	0	0
1	0	1	1	1		1	0	1	0	1	1
1	1	0	1	1		1	1	0	1	0	1
1	1	1	1	1		1	1	1	1	1	1

$A(B+C)$　　　　　　　　　　　　$AB+AC$

Fig. 8.60.　$A(B + C) = A.B + A.C.$

A	B	A+B	$\overline{A+B}$
0	0	0	1
0	1	1	0
1	0	1	0
1	1	1	0

$$\overline{A+B}$$

A	B	$\overline{A}$	$\overline{B}$	$\overline{A}.\overline{B}$
0	0	1	1	1
0	1	1	0	0
1	0	0	1	0
1	1	0	0	0

$$\overline{A}.\overline{B}$$

Fig. 8.61. $A + B$ $A.B$.

A	B	AB	$\overline{AB}$
0	0	0	1
0	1	0	1
1	0	0	1
1	1	1	0

$$\overline{AB}$$

A	B	$\overline{A}$	$\overline{B}$	$\overline{A}+\overline{B}$
0	0	1	1	1
0	1	1	0	1
1	0	0	1	1
1	1	0	0	0

$$\overline{A}+\overline{B}$$

Fig. 8.62. $A.B$ $A + B$.

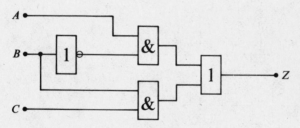

Fig. 8.63.

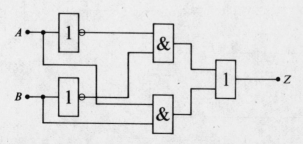

Fig. 8.64.

8.B   **1.** See text.   **2.** Output $= 20/21 \times 12 = 11 \cdot 4\,\text{V}$.
3. $12\,\text{V} - 20\% = 9 \cdot 6\,\text{V}$
If load is R

$$\frac{9 \cdot 6}{R} = \frac{12}{R+1} \quad \text{giving } R = 4\,\text{k}\Omega.$$

Hence maximum number is 5 (answer b).
4. See text.   **5.** See Fig. 8.65.   **6.** See text.
7. $Z = A \cdot \bar{B} + \bar{A} \cdot B$. See Figs. 8.66 and 8.67.
8. $Z = A \cdot \bar{B} + B \cdot C$
(a) See Figs. 8.68 and 8.69.
(b) See Figs. 8.70 and 8.71.
9. $Z = A \cdot B + \bar{A} \cdot \bar{B}$
(a) See Figs. 8.72 and 8.73.
(b) See Figs. 8.74 and 8.75.
10. Adding extra NANDs gives the diagram shown in Fig. 8.76.
Obviously $Z = A + \bar{B} \cdot C$.

A	B	C	$\overline{ABC}$	$\overline{A}+\overline{B}+\overline{C}$
0	0	0	1	1
0	0	1	1	0
0	1	0	1	0
0	1	1	1	0
1	0	0	1	0
1	0	1	1	0
1	1	0	1	0
1	1	1	0	0

Fig. 8.65.

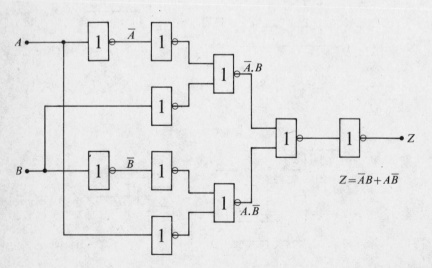

$Z = \bar{A}B + A\bar{B}$

Fig. 8.66.

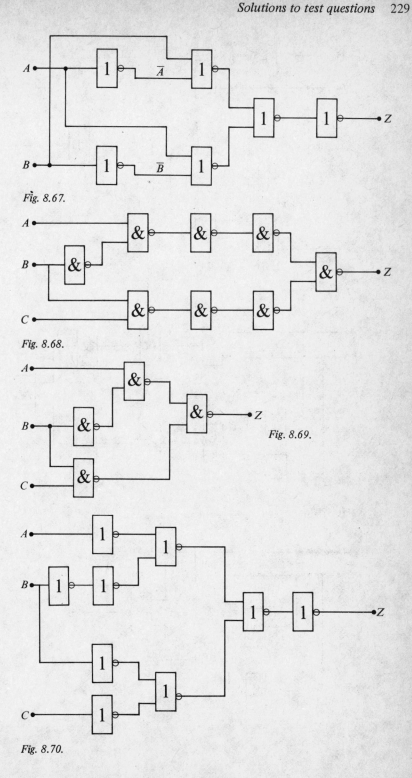

Fig. 8.67.

Fig. 8.68.

Fig. 8.69.

Fig. 8.70.

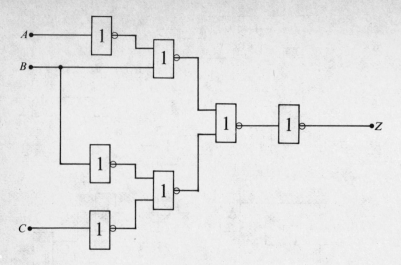

Fig. 8.71.

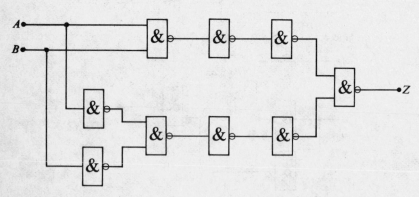

Fig. 8.72.

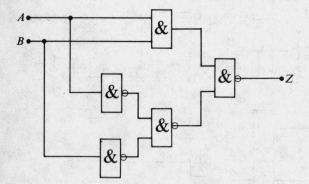

Fig. 8.73.

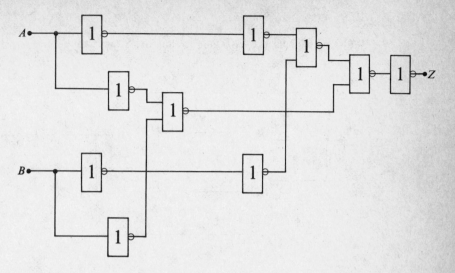

Fig. 8.74.

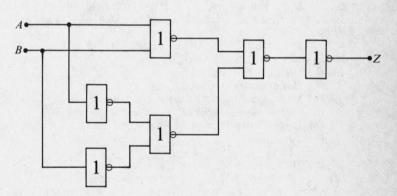

Fig. 8.75.

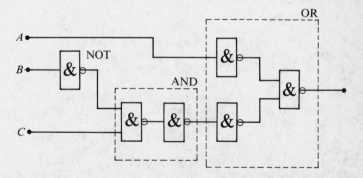

Fig. 8.76.

8.C   **1.** See text.   **2.** See text.
3. See Fig. 8.77.
If $Q = 1$ it opens the AND gate feeding the R input and resets the bistable.

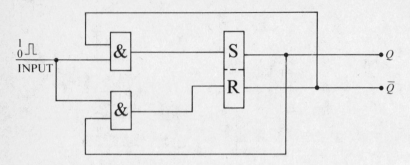

Fig. 8.77.

9.A   **1.** c.   **2.** Thermionic, radiation, secondary, field.
3. Tungsten (least efficient), thoriated tungsten, barium and stronium oxides.
4. b.   **5.** a.   **6.** See Fig. 9.4.   **7.** See Fig. 9.5.   **8.** a.   **9.** c.
10. (a) Reverse current is zero.
    (b) Higher volts drop.
    (c) Heater current required.
    (d) Mechanically weaker than semiconductor.
    (e) Electrically more robust than semiconductor.
    (f) Larger than semiconductor for given rating.
11. (a) 10·5 mA
    (b) 72 V
    (c) 78 V
12. a.
13. Change of anode voltage divided by the corresponding change in anode current.
14. a.c. 2·9 kΩ
    d.c. 6·7 kΩ.
15. See section 9.3.

9.B   **1.** b.   **2.** See eqn (9.1).   **3.** See Fig. 9.7.
4. See Fig. 9.8 and Fig. 9.9.   **5.** a.   **6.** 4·7 kΩ.   **7.** See eqn (9.3).
8. b.   **9.** a.   **10.** c.   **11.** See section 9.7.   **12.** a.
13. See Fig. 9.18.   **14.** See text.
15. 15 mA. Normal load-line construction using the $V_{gc} = 0$ V curve.
16. 8 mA. As problem 15 but using the $V_{gc} = -2$ V curve.
17. b; 400 V/2 mA = 200 kΩ.
18. b; V_{ak} swings from 290 V to 75 V = 215 V peak to peak.
Gain = 215/3 = 72.
19. 1·75 mA; 225 V.   **20.** 400 V × 1·75 mA = 700 mW.
21. Load line vertical. 2·15 mA.
22. Current = (3·1 − 1·5) mA = 1·6 mA peak to peak; voltage zero.

9.C   1. See Fig. 9.19.   2. See text.   3. b.   4. a.   5. See Fig. 9.21.
6. a.   7. b.   8. See text.
9. As the spot tries to move both horizontally and vertically at the same time it will produce a straight line on the screen. This will be at 45° if the sensitivities of the X- and Y-plates are the same.
10. Similar to 9 but one waveform is at its maximum when the other passes through zero. If the plate sensitivities are equal a circle will be produced.

Index